A LEVEL
Questions and Answers

GEOGRAPHY

Penny Goodman & Jonathan Hughes

SERIES EDITOR: BOB McDUELL

Contents

INTRODUCTION

The importance of using questions for revision — 1

Types of question — 1

Geographical skills and assessment objectives — 7

REVISION SUMMARIES AND QUESTIONS

Physical Geography — 9

1 The atmosphere — 9

2 The biosphere and ecosystems — 14

3 Hydrology and drainage basins — 21

4 Arid and semi-arid environments — 27

5 Coastal environments — 30

6 Glacial and periglacial environments — 34

7 The lithosphere and plate tectonics — 37

Human Geography — 41

8 Settlement — 41

9 Population and resources — 49

10 Industry — 55

11 Agriculture and food — 60

12 Tourism and leisure — 64

13 Spatial inequality, regional disparities and development — 67

Practical Geography — 72

ANSWERS — 73

Introduction

THE IMPORTANCE OF USING QUESTIONS FOR REVISION

Past examination questions play an important part in revising for examinations. However, it is important not to start practising questions too early. Nothing can be more disheartening than trying to answer a question which you do not understand because you have not mastered the topic. Therefore, it is important to have studied a topic thoroughly before attempting any questions on it.

How can past examination questions provide a way of preparing for the examination? It is unlikely that any question you try will appear in exactly the same form on the papers you are going to take. However, the examiner is restricted on what he can set, as questions must cover the whole syllabus and test certain Assessment Objectives. The number of totally original questions you can set on any part of the syllabus is very limited and so similar ideas occur over and over again. It certainly will help you if the question you are trying to answer in an examination is familiar and you know you have done similar questions before. This is a great boost for your confidence and confidence is what is required for examination success.

Practising examination questions will also highlight gaps in your knowledge and understanding which you can go back and revise more thoroughly. It will also indicate which sorts of questions you can do well and which, if there is a choice of questions, you should avoid.

Attempting past questions will familiarise you with the type of language used in questions.

Finally, having access to answers, as you do in this book, will enable you to see clearly what is required by the examiner, how best to answer each question, and the amount of detail required. Attention to detail is a key aspect of achieving success at A level.

TYPES OF QUESTION

Whatever the type of question, the most common error that students make is to throw away marks by not reading the instructions and/or the question. It is essential to check the rubric (the instructions on the front page of the exam paper). How many questions must I answer? Are there constraints on which sections I can answer from? Every year able candidates lose marks through these mistakes, do not be one of them!

There are four kinds of question in geography examinations: structured questions, essays, applied and practical ones and decision-making exercises. It is essential that you look at past papers, and the questions in this book, to find which types are set by your board. In addition some syllabuses require a compulsory project, and some boards offer this as an alternative to a written paper. The approach to the investigative study is not covered by this guide.

Question selection

Unless you are dealing with a paper or section that is compulsory, the first (and a very important) stage in answering the paper is to choose your questions. Too often, candidates when leaving the exam room or during the course of the exam realise that there were questions that they could have answered better. You must get it right first time, which means devoting up to 5 minutes selecting the questions you can answer best in terms of your knowledge and understanding.

❶ Scan the entire question paper.

❷ Disregard non-compulsory sections for which you have not been prepared. Why try to answer a question that you have no knowledge of?

❸ Identify possible questions you could answer, by putting a question mark in the margin. Consider both the content required by the question, and the phrasing. Are you in a position to discuss the topic? Do you have enough case study material for 'With reference to examples you have studied, consider the statement that...'?

Introduction

❹ Select the questions that you think that you can answer best, making sure that you select the right number and obey the rubric instructions. If you have written a similar essay in the past or on a similar topic this is probably a good bet, but *never* fall into the trap of assuming that it is identical to the one you did before! A good 'A' level answer will focus on the question, not be vaguely relevant to the topic set.

❺ Avoid selecting a question in order to be 'different'. It is better to answer a stock question (perhaps one that your teacher has 'spotted') **well** than a more obscure one that you have not been prepared for, and do not have enough information for.

Short structured questions

Aim: The examiners are testing your understanding and knowledge of geographical concepts, principles, and theories.

Background: To do this you are generally provided with data, simple maps, and photographs which require geographical skills in their interpretation, judgement and evaluation. These types of question are usually compulsory and test the core topic material.

For Physical Geography this is almost always:
    The atmosphere
    The biosphere and ecosystems
    Hydrology and drainage basins

For Human Geography the core material tends to be:
    Population and resource relationships
    Processes of urbanisation and industrialisation

How to answer this type of question:

1 Check how long you have for this paper. Set time limits for each question according to question length, so for 5 evenly weighted questions in 1 hour 15 minutes you have 15 minutes a question. Don't overrun, especially if you have an essay paper to follow.

2 Read the question *carefully*. What is the question asking you to do? '*Describe*', '*evaluate*', '*explain*'. Also, be sure to pay close attention to terminology. Is the question about develop**ed** or develop**ing** countries, or about **inter** or **intra** city shopping patterns? Use the information that is provided, it is there for a reason, to focus you on the topic and as a basis for some answers, it is not just there to make the examiner's questions more appealing. Study all the map, diagram, and photographic information given to you.

3 The first parts of the question are often recall. For example, '*What is the weathering process that can attack a rock that contains feldspar and quartz?*'

Or, it may test the application of basic principles such as, '*Which of these climates is likely to have the fastest rate of chemical weathering?*'

4 Answer these questions succinctly (clearly and concisely) in a straightforward sentence or single word. You must confine your answer to the lines provided. The number of lines and the number of marks awarded is a good indication of how much is required. As a guide, 1 mark probably only needs a simple sentence or even one word. If 2 marks are being awarded then it is likely that an explanation is required as well as a statement. If you can't fit your answer in, or it is far too short, consider whether you have the right answer!

5 Sometimes you may think that you know nothing about the subject being asked. For example, a question might ask how '*shakeholes*' are formed. A 'shakehole' is a dialect word (which is not common and one that you may not know) for a depression that forms in the soil above limestone rock. If you panic at this stage, which some students may do, and go on to the next question you will lose marks. In fact a thorough look at the information given in the question would show that these depressions form where the joint density is greatest and therefore most vulnerable to solution. THINK, STAY CALM and DO NOT PANIC.

Introduction

6 If you really don't understand a question, do not waste time agonising over it, but go on to the next one.

7 Go back at the end of the exam to any question you have missed out and try again, never leave blanks (see point 6). Check your answers for silly errors which may have changed the meaning of your response (for example 'do' rather than 'do not').

Essays: Extended prose

Aim: The examiners want to see if you understand the topics and whether you can apply your knowledge about the subject. You are also being assessed on your ability to organise and present ideas in coherent extended prose. It is important that you include detailed case studies. The questions are set in such a fashion that you are able to select your own material. For example, *'With reference to developing countries that you have studied, discuss the problems arising from rural to urban migration'* does not tie you to any named country. Marks will be awarded according to the degree of knowledge and understanding shown. More importantly your ability to answer the question and evaluate your evidence is tested.

Background: You usually have 45 minutes per question, but check your rubric, and note that the material examined often includes optional themes. You must practise this type of question, as it is often the one candidates are weakest at (GCSE does not require essay writing).

There are two main types of essay. The first type is divided into a couple of sections with the first part requiring the definition of some basic terminology, or description of a feature or figures. This is followed by the part of the question which requires explanation or evaluation; this part carries more marks. The second type of question tends to consist of a statement, which you are required to support using case study material, or which needs to be discussed.

How to answer this type of question:

1 Once you have narrowed down your choice it is essential to spend at least 5 minutes planning your essay. This may seem a waste of time, but it is the most crucial period. Making an error at this stage will cost you many marks and even several grades. You must answer the question. It is no good just writing down things that are vaguely related to the question title. To make sure you answer the question, follow these guidelines.

On a piece of paper write out the question title, then carefully underline the key words. This includes words that tell you what to do, for example, *'compare'*, *'explain'*, *'account for'*, *'discuss'* and less often at 'A' level *'describe'*. You also need to identify the subject matter such as *'less developed'* or *'climatic types'*. For example:

'(a) <u>Explain</u> the nature of <u>constructive</u> and <u>destructive</u> waves. [9]

(b) <u>Explain</u> the <u>formation</u> of <u>spits, bars and tombolos</u>. By use of <u>examples</u> demonstrate <u>why</u> and <u>how</u> these features can be subjected to <u>rapid change</u> through both <u>human and natural agencies</u>. [16]'
UCLES

Make sure you really are careful here, it is too easy to misread a question if you are worried or nervous. The example below shows what can result:

'Explain how processes, other than river action, may modify the landforms of a drainage basin.'
WJEC

If you were *expecting* a question on rivers it would be easy to think that this was it, but of course it deals with everything *but* rivers!

2 Quickly define any terminology that you are asked to, or describe the features that are required. Even if the question does not specifically require definitions it is still worth briefly outlining geographical terms that appear in the question.

3

Introduction

3   Having identified the subject matter, think about the key influences, processes and/or factors that will control the feature, pattern, and/or trend. Write these down as headings. These will form the paragraphs and therefore determine the structure of your essay. Regard your plan as a checklist of points only. It is not a rough draft of your essay.

4   For each of these paragraphs you need to think of good case study material to back up your points. In an essay it is not enough to be able to recall the points, you must be able to string the points into a logical sequence of ideas, with each point being justified. In point form add these to your headings.

5   In summary:
    THINK about the question asked
    RECALL the relevant information
    APPLY your knowledge to the question, showing your
    UNDERSTANDING of the issues
Only then can you begin to write the essay.

6   If the essay is divided into two parts make sure you don't spend too long on the first straightforward part (look at the mark allocation). Description does not earn many marks. Answer briefly and accurately, quoting figures from the paper where appropriate.

7   **The opening paragraph is the key to the essay**. You can use it in three ways. First, you can use it to define any terminology, and to make a simple statement about the factors that affect the subject in question. This is a business-like way of proceeding, and has the advantage of informing the examiner that you know about this topic and how you intend to proceed.

Second, you can open dramatically with a strong case study or statement which grabs the examiner's attention, but illustrates the general points you wish to make. This can be brilliantly successful, but is more risky, especially if it is not followed up by an essay with a tight structure. If you adopt this method you really need to know your subject. It is safer and far easier to stick with with the tried and trusted idea of telling the examiner what you are going to say in your introduction. Then tell him your answer in the body of the essay, and in the final paragraph tell him that you have told him!

Third, you can begin your essay with a quote, which is relevant to the topic. For example, if you were answering a question on public transport you could use:

'*Progress be dammed. All this will do is to allow the lower classes to move around unnecessarily...*' The Duke of Wellington, on seeing the first train.

This introduces the topic in an unusual manner, and allows you to comment on this particular view.

8   **Then the main body of the essay.** Deal with each of your points from your essay plan, backing them up with examples. Examples may range from single word references such as '*Many declining industrial regions in the U.K. e.g. **Tyneside**'* to substantial case study material. Examples should not be vague or hazy. It isn't enough to state that '*birth rates are high in Africa*'! (Africa is a large continent and, of course, displays huge variations in birth rates.) Where possible quote specific examples, and give figures to expand and illustrate your point. If appropriate, expand the case study material pointing out that the situation is not always as simple as it would appear.

Where possible use **local examples**; these are different from the standard examples in text books, and you may well be able to think of these off the cuff!

9   A surprising number of candidates do not seem to realise the value of **diagrams** and **simple sketch maps**. Use diagrams as often as possible. They are of great value because:

   • They can summarise, and show the features of, for example, a soil more clearly than you can describe it in words.

Introduction

- They save time. It is often faster to sketch something than describe it.
- Using flow diagrams or systems diagrams enables you to show quite complex interrelationships, in an easily comprehensible way. e.g. Myrdal's idea of cumulative causation.

    Do not worry if you are not an artist. Adequate labelling will make a poor diagram good. NEVERTHELESS a diagram is of little value unless it is clear, labelled accurately, has a title and is integrated into the text. Do not copy out a diagram that is already on the exam paper, this is an utter waste of time! Look at diagrams included in this book to see what is required.

10 Exam questions often deal with geographical models and theories; these are simplifications of reality, so try to avoid a blind acceptance of them. Show the examiners that you are aware of the limitations of models and theories.

11 **The final paragraph**. This is important. You may simply need to summarise the information that you have already communicated to the examiners, but more often the final paragraph has a more important function. If the title says:

   'Assess the relative importance of ...'

   or

   'To what extent do you agree with the view that ...'

   or

   'Compare the development of ...'

   you need to use the final paragraph to come up with some sort of conclusion. You need to explain which things you regard as being most the most important, or whether on balance you agree with the statement given.

12 Do not run over your time allocation, because diminishing returns set in and the more time you spend on one question the less you have for another. It is no good getting 21/25 for one question but only 5/25 for another. This 'rule' applies even when you think you don't know very much about your final question. It is very easy to pick up 10 marks, but very rare to get over 23/25.

13 If you are about to run out of time, answer in point form and refer the examiner to your plan.

Applied and practical papers

Aim: Here your ability to apply your knowledge, and understanding of skills and techniques is tested by making you consider *unfamiliar* geographical situations and problems. In this type of question you are often required to explain how data is collected and should be best represented and analysed. The questions tend to focus on the core subjects. Questions may be based on O.S. maps, aerial and satellite photographs, field work and statistical data, and land-use maps.

Marks will be awarded according to the degree of understanding, evaluation and level of skills demonstrated.

Background: To answer this type of question you must be familiar with sampling methods, statistical analysis, and techniques for representing statistical data, for example graphs, charts and maps. Revise these thoroughly!

How to answer this type of question:

1 Read through the questions, deciding which you can answer best. To do this, look at the mark allocation; it is not worth rejecting a question merely because you are unable to do a small part. The best questions to answer are those that you have first hand experience of, perhaps on a field trip. This is because you will have gone through the relevant stages and it is easier to remember things that you have done.

Introduction

2 Plan your answers as outlined in the essay section. You must focus this time on the geographical techniques required to solve the problem; the examiners do not want to know what you would expect to find, neither do they (usually) require you to construct diagrams. When illustrating the points you make, you need to quote directly from the maps and figures set. Give grid references as well as place names.

3 Make sure you answer the question set and that you have taken the mark scheme into account.

4 Remember, the examiners want to see how able you are at applying principles. For example, when sampling, you need to select a sample that is representative of the total population being studied. However, you also need to demonstrate that you are aware of any constraints that may operate. An example might be a question that deals with the difficulties in conducting a shopping questionnaire. You need to discuss the sample size required, the influence of the time of day and week, the effect standing in different locations might have, and the problem in getting a good response rate from different people, etc.

5 Your answer must demonstrate an understanding of the selection and detailed knowledge of the construction of diagrams. For example, when explaining and using choropleth maps, it is rarely sufficient to define simply what these are. You may also need to discuss in detail the means of determining both the number of classes chosen and the class boundaries.

Decision-making exercises

Aim: At first, decision-making exercises appear very different from any other examination questions, but the general principles are the same. This type of exercise is designed to assess your ability to undertake a geographical enquiry, but it differs from the traditional data-response exercise outlined above as these exercises require you to consider peoples' values, attitudes and views. The examiners are testing your ability to digest and organise information, and whether you understand this information. Finally, they wish to know if you can write extended prose, demonstrating graphical and cartographic skills, in a coherent and structured manner.

Background: Decision-making exercises are a specific type of structured essay question. The case study material is likely to be new to you, but the topic areas (e.g. an Environmental Impact Assessment) and the techniques should be familiar.

Therefore, you have to condense the 'learning experience' into hours rather than weeks. This is a very important point; you have to make decisions quickly, which is a test in itself. For example, what information do I use and what do I discard? How do I present my report and what recommendations am I going to make?

A typical question might be to investigate the route taken by a new motorway, from the viewpoint of a planning inspector. This question is an essential component of the 16-19 Geography project. The exercise takes two and a half hours.

How to answer this type of question:

1 Read and identify the key instructions given to you, that is, underline the important words, etc. Use these to produce a checklist of tasks.

2 Allocate your time carefully. If marks are shown for each section use this as a guideline as to how much of your time a certain section needs. If a breakdown is not given, you have to decide which areas are more important and thus warrant more time.

3 Using your checklist identified in **1** above, read the information and resources provided. The information that you will need to consider will be in the form of maps, statistics, articles and reports. Underline those facts and figures, quotes, and titles that are relevant. Do *not* underline everything; if you do, nothing becomes important as everything has been upgraded. Even if you have prior knowledge of the topic area, do not rely on this. The information provided has to be the basis of most of your report.

4 Now go through the material once more, targeting and highlighting those points needed to answer this part of the Decision-Making Exercise. Using this highlighted material, plan your answer under headings in point form on a piece of rough paper.

Introduction

5 In the exercise, you are often asked to assume a role. Do not be put off by this, and do not act! This role is designed to help you assess the material from a neutral perspective, in an unbiased and impartial way. This is crucial; these questions are about objective assessment, not your personal emotions. You must use balanced reasoning and take into account the views of opposing groups which will not necessarily reflect your own perspectives, values or beliefs. However, a decision does have to be made, so if the evidence supports one argument do not be afraid to give it your support.

6 You may also be required to offer your own considered views in the Decision-Making Exercise and to decide upon your own personal position.

GEOGRAPHICAL SKILLS AND ASSESSMENT OBJECTIVES

As an A-level candidate you need to meet the criteria that the examiners are looking for. All the geography syllabuses have as their aim the intention that you, the geography student, should demonstrate ability in the following four areas: knowledge, understanding, skills, and values and decision-making.

Knowledge

Knowledge is the basic and, arguably, the most important building block. Without reasonable knowledge and grasp of the subject, you cannot answer any question properly. Without facts and figures to hand you waste your time in the examination because you are struggling to remember information, and without this you lose structure, which is vital when writing essays or longer response answers. Knowledge can be divided into the following areas:

❶ **terminology** e.g. chernozem, hierarchy, ecosystem

❷ **processes** e.g. corrasion, counterurbanisation, migration

❸ **places** all geographical things happen at specific locations, therefore you have to have examples of places

❹ **theories and principles** e.g. Demographic Transition Model, Christaller's Central Place.

It is essential that you revise and learn examples and facts.

Understanding

However, once this database of knowledge has been acquired, a good candidate needs to be able to understand the ideas and facts, and be able to apply them. Without understanding you can only achieve a low grade.

There are four key areas of understanding:

❶ understanding of the **processes** that determine physical and human geography, for example, you need to be able to explain the migration of people in terms of economic, social and political factors.

❷ **models and theories**, for example the evolution of a depression (low pressure system) in meteorology. Models are valuable because you can generalise and predict from them.

❸ **temporal and spatial change** is the concept that patterns of distribution, processes and characteristics will vary over time and space. For example, the process of urbanisation versus counterurbanisation changes as development proceeds.

❹ **inter-relationships** between the physical environment and human activity. For example, those between soil type and farming.

Introduction

Skills

Examiners will be looking for your ability to use different techniques in a geographical context. This is important whether you are doing essays, structured questions or decision-making exercises. You must be able to interpret data given to you, plan and communicate your answer in a coherent manner using the following skills:

❶ **communication** – you should be able to show the examiner that you can think logically and present a structured answer (by writing, cartography, diagrams and graphs).

❷ **organisation** – you should be able to select, collect, and organise geographical data. It is vital to make accurate and objective observations.

❸ **problem analysis** – you must be able to evaluate and interpret data from a variety of sources to solve problems.

Values and decision-making

Finally, examiners want you to recognise that your own perspective is only one of many, and that different cultures view the same facts through different eyes:

❶ **cultural diversity** – you should appreciate that the values and beliefs of other cultures, religions, traditions and economic and political systems may have an impact on the geography, for example in attitudes towards birth control.

❷ **the environment** – you should be aware of different attitudes towards environmental issues.

❸ **scenarios** – there may be different opportunities and constraints in countries, for example inside and outside the EU.

❹ **perceptions** – these affect decision-making within a geographical context. People act on the basis of their perceptions and limited information fields, which they treat as fact, even though this may be a distortion of reality.

The atmosphere 1

What is the difference between meteorology and climatology?
Weather can be regarded as the atmospheric conditions operating at any particular time and place. The study of weather is known as **meteorology**. **Climate** may be regarded as the average characteristics of wind, precipitation, air pressure, humidity, and temperature collected over a long period of time (30–50 years); climate is therefore 'accumulated weather'.

The structure of the atmosphere and the atmospheric circulatory system
The earth's atmosphere consists of four layers, each with their own characteristic temperatures and gases. The lowest layer is the **troposphere**, about 8–16 km deep (thinnest near the poles), and it is this 'envelope' that contains the weather. The troposphere is separated from the next layer, the **stratosphere**, by the **tropopause**. The stratosphere contains the **ozone** layer (O_3) important in absorbing and filtering ultraviolet radiation. The phrase 'ozone hole' refers to the reduction of ozone within the stratosphere principally over the poles caused by the increased anthropogenetic emission of CFCs. The ozone layer reduces the harmful ultraviolet rays; any thinning will mean more radiation enters the lower atmosphere.

The heat budget and the atmospheric circulatory system
Global variations in the **heat budget** are caused by spatial and temporal (time) variations in **insolation**. The **Coriolis force** combined with the variations in the heat budget produce vertical and horizontal movements in the atmosphere as excess heat is transferred from the Equator towards the cooler poles. There is a surplus of energy at the Equator due to the altitude of the sun in the sky. This is because at the poles incoming radiation has to heat a larger area of the earth's surface than in lower latitudes and more is absorbed in passing through the wider atmospheric envelope at the poles. Heat transfer is accomplished by radiation, conduction, convection and latent heat. Atmospheric circulation is broken into three cells within each hemisphere, (the **Hadley**, the **Ferrel** and the **Polar** cells). The tricellular nature of the earth's atmospheric system produces some of the key features of the global climates. The rising air mass over the Equator is associated with high levels of convective precipitation, whereas subsiding stable air produces a high pressure zone in the region 30°N or S causing desert conditions, and an area of instability in mid latitudes where polar air meets warm air from the subtropics.

Local variations in heating are produced by height above sea level, proximity to the sea, ocean currents and prevailing winds, and are also seasonal, due to changes in the length of day, aspect, cloud cover and land surface.

Water vapour in the atmosphere
Humidity measures the amount of water vapour in the atmosphere. Two measures are important: **absolute humidity**, the mass of water in g/m^3, and **relative humidity**, which is the quantity of water vapour in the air compared with the maximum amount that could be held in the air at a specific temperature, expressed as a percentage. At 100% the air is saturated. Warm air is able to hold more water vapour than cold. Further cooling of saturated air results in condensation when the **dew point temperature** is reached. Condensation results from **radiation cooling**, **advection cooling** (due to horizontal movement of air), vertical uplift due to fronts and mountains (**orographic cooling**) and **convective /adiabatic cooling**. The last is very important. Look at the question and answer section for diagrams and on **lapse rates** which are an essential part of the course. Refer to these for definitions on **environmental lapse rates (ELR)**, **dry adiabatic lapse rates (DALR)**, **saturated lapse rates (SALR)**, **air stability** and **instability**. Remember, air is stable when the SALR and DALR lie to the left of the ELR, because the mass of air is cooler, and therefore denser than the surrounding air.

Condensation results in **precipitation**, and you should expect questions on rainfall – frontal, orographic (relief), and convectional as well as **snow**, **fog** and **hail**. Fogs result from radiation cooling and advection; smogs result when atmospheric pollution combines with fog. It is particularly noticeable when urban areas lie in valleys. **Temperature inversions** (where warm air

1 The atmosphere

REVISION SUMMARY

overlies cold air), often associated with anticyclonic conditions, may allow smogs to persist. As the air is stable and ground heating is reduced, the sun cannot penetrate the fog to warm the air.

Weather systems

Countries in mid latitudes such as the UK are influenced by several air masses, which cause considerable variations in the weather experienced. In contrast, equatorial areas experience less variation as fewer air masses are involved. **Air masses** are fairly homogeneous bodies of air that derive their characteristics of temperature and humidity from their source regions. Air masses may be modified by the land or seas they pass over, becoming stable or unstable in the process. The British Isles may be influenced by Arctic maritime (Am), Polar maritime (Pm), Polar continental (Pc), Tropical maritime (Tm), and Tropical continental (Tc) air masses. The boundary where warm and cold air meet is termed a **front**, and **depressions** form along these. Depressions are low pressure systems within which winds blow anticlockwise in the northern hemisphere. It is important to be able to recall the processes involved in the formation and the decay of depressions, and the weather associated with their passage. These can be checked with the diagrams in the question section on atmosphere. Britain is also affected by **anticyclones** (high pressure systems) in which air is subsiding. The weather associated with these varies according to the season. In winter cold, clear weather with fogs forming at night can be expected, whereas in summer hot, sunny weather often terminated by thunderstorms is typical.

For low latitudes there are two main weather systems on the syllabuses; monsoons and tropical cyclones (hurricanes and typhoons) the latter resulting from intense low pressure systems. You should again be familiar with the climatic factors influencing their formation and their associated weather, particularly wind patterns and air pressure.

Microclimates

A microclimate is defined as the study of local climate found within a small area and a few metres from the ground, which is determined by local factors rather than the regional weather systems. The key here is to explain variations in humidity, air and soil temperatures, and wind characteristics in terms of differences in albedo, aspect, and shelter (such as buildings and trees) as well as proximity to water. **Albedo** is a measure of the radiation reflected from a surface compared with the incoming radiation, expressed as a percentage or ratio. Snow has an albedo of 85%, but concrete only 17–27%. Most syllabuses expect knowledge of anabatic and katabatic winds.

Climatic hazards, weather forecasting and economic activity

Parts of the question often require an assessment of the impact and effect of weather on human activity. Questions on hurricanes, blizzards and local winds (e.g. chinook or Föhn) are set. The utility of forecasting such hazards with a view to ameliorating (lessening) the effects should be known. Phenomena such as the jet stream, frost pockets, air quality, and areas susceptible to fogs, all have implications for people such as pilots, farmers, car drivers, as well as being a health hazard. Climate and weather are important for all farmers, as it directly affects their profitability, but in developing countries, for example those within the Sahel and the Indian subcontinent, climatic variation may result in famine.

Global climatic issues

Three issues frequently occur. First, the impact volcanic eruptions and human activity such as deforestation or the increase in CO_2 from burning fossil fuels have in producing the 'greenhouse effect'. Second, changes to the ozone layer, whether resulting from natural aerosols, such as volcanoes, or human activity which result in potential health hazards and damage to food chains. Finally, the effects of urbanisation on localised heating and acid rainfall. It is essential that any account you give is balanced, and discusses the reliability of the data and alternative causes of change rather than being a sensational account based on media hype.

If you need to revise this subject more thoroughly, see the relevant topics in the Letts A level Geography Study Guide.

The atmosphere 1

1 Fig. 1 shows unstable atmospheric conditions.

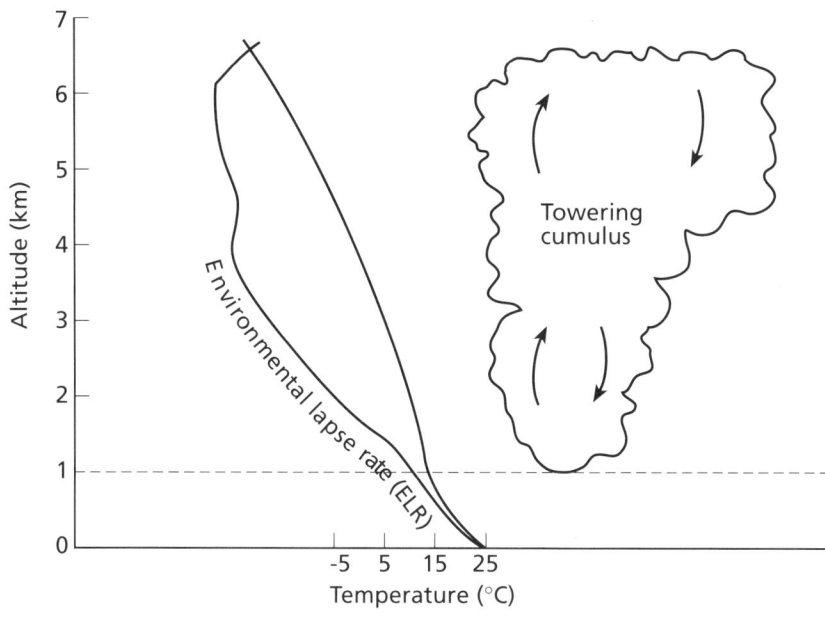

Fig. 1

(a) Complete the diagram by labelling the following features:

(i) dry adiabatic lapse rate   DALR;
(ii) saturated adiabatic lapse rate SALR;
(iii) condensation level   CL. (3)

(b) What is meant by the environmental lapse rate (ELR) shown on the diagram? (2)

(c) Explain why the saturated lapse rate (SALR) is at a different rate from the dry adiabatic lapse rate (DALR). (2)

(d) (i) Describe the weather conditions normally associated with unstable air conditions. (4)

(ii) Areas reliant upon tourism benefit from stable air conditions.
Suggest **two** reasons why this is so. (4)

AEB

2 Account for three of the following:

advection fog; jet stream; hail; anabatic (up-valley) wind; orographic rainfall. (9,8,8)
ULEAC

3 (a) (i) Define the terms:

dry adiabatic lapse rate;
saturated adiabatic lapse rate. (4)

(ii) Briefly outline why an understanding of these terms is important for weather forecasting. (3)

1 The atmosphere

QUESTIONS

(b) Fig. 2 shows a synoptic chart for 06.00 hours on 15 May 1979.

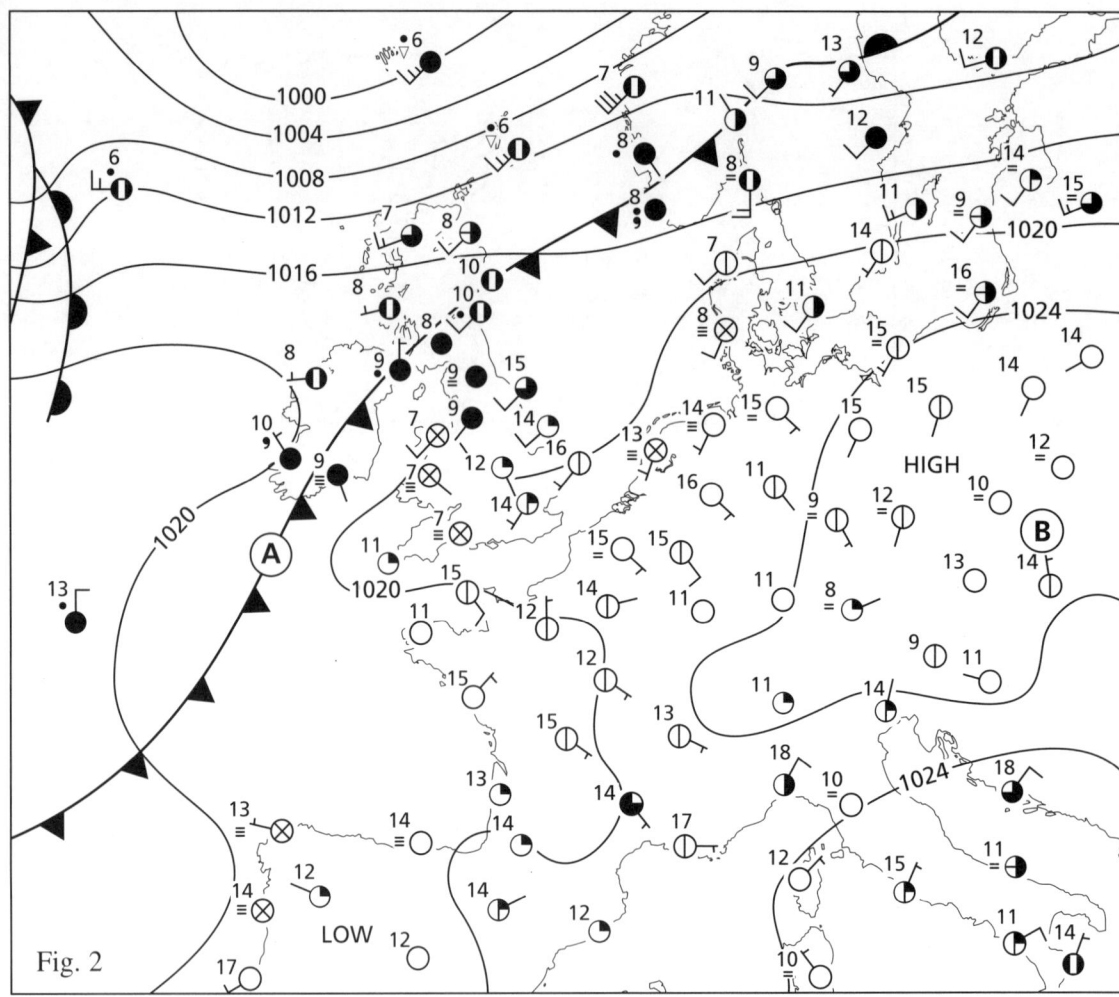

Fig. 2

(i) Identify the weather features marked **A** and **B** on the chart. (2)

(ii) Interpret the different weather conditions being experienced in northern and southern Britain. (6)

(c) Explain how mists and fogs are produced. To what extent have human activities affected the incidence and distribution of fogs? (10)

UCLES

4 Study Fig. 3, which shows sea surface temperatures and tracks of tropical cyclones.

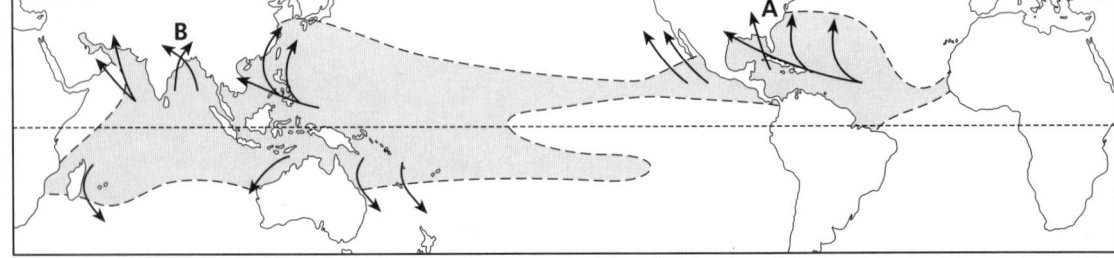

Fig. 3

Key
→ Main tropical cyclone paths
----- Maximum sea surface isotherm 26.5°C
▒ Sea area with maximum temperature

12

The atmosphere 1

(a) Account for the tracks of tropical cyclones shown in Fig. 3. (6)

(b) For tropical cyclones, outline:

    (i) the main meteorological hazards; (6)

    (ii) the other associated hazards. (5)

(c) For areas A and B on Fig. 3, explain the differing impacts of tropical cyclones on human activity. (8)

ULEAC

5 Examine the ways in which different types of vegetation may create their own micro-climate. (25)

Oxford & Cambridge (Specimen)

6 Describe and account for the distinctive features of urban climates. (25)

ULEAC

7 Explain what might be learnt about weather and climate from meteorological data collected at a school or college. (20)

WJEC

8 (a) Insolation is the energy reaching the earth's atmosphere from the sun. Discuss ways in which the atmosphere affects the amount of insolation reaching the earth's surface. (7)

(b) Discuss the role of the following factors in the heating of the atmosphere:

    (i) specific heat capacities of land and sea;

    (ii) albedo of different types of surface;

    (iii) altitude;

    (iv) aspect. (11)

(c) Describe how, and explain why, heat energy is transferred from one latitude to another. (7)

NEAB

9 Why does it rain? (25)

ULEAC

2 The biosphere and ecosystems

REVISION SUMMARY

This is a core topic. The syllabuses focus on two main aspects. First, the principal factors which govern the structure, function, and distribution of biomes, and second the modification, manipulation and change caused by human economic activities.

What are ecosystems and biomes?

An **ecosystem** is a functioning and interacting system of living organisms and the physical environment within which they live. It consists of an ordered and interacting community of plants and animals and the abiotic (non-living) environment which includes - climate, geology, relief and soil. Ecosystems vary in size from a rotting log to a **biome**. A biome is a climax community of plants and animals found within a climatic region, such as the savanna or rain forest.

Ecosystems are made up of interlinked parts, and rely on energy inputs from the sun. The energy allows plants and bacteria (**primary producers or autotrophs**) to photosynthesise, and produce **biomass** (the mass of living matter). The plants may be used by animal **consumers** above them in the food chain for energy, and these **primary consumers** may in turn be consumed by carnivorous consumers. This produces a **food chain or web**, each stage being referred to as a **trophic or energy level**, with plants as the first trophic level.

A trophic pyramid showing reduced biomass at each trophic level for a Florida river

There are two basic processes in an ecosystem, the first is the flow of energy and the second is the flow of nutrients in the system. Part of the sun's energy is converted to plant material, and a proportion of this is then transferred to other levels by consumption. The system is not, however, efficient, and at each stage there are losses of energy because some energy is needed for activities such as respiration and movement. This imposes limits on the numbers of species present and means most systems are limited to four trophic levels. Ecosystems are open systems, nutrients enter the system by the decay of rocks, and additions from volcanic ash and rain water. Losses occur through **leaching** or harvesting of crops for consumption outside the area. Material is cycled through the system by the activities of the **decomposers** such as fungi. Diagrams showing the cycling and storage of nutrients in different biomes often form the basis of questions (see Question 4). These diagrams were devised by Gersmehl (see Figs. 4 and 5).

Global distributions of biomes

Biomes are named after the vegetation found in them, for example coniferous (**boreal**) forest, but also refer to the animals and soils found in them. Recognisable vegetation zones exist because of the influence of climates in regulating plant growth. Four main factors control the distribution of biomes: climatic, topographic, **edaphic** (soil variations often caused by rock type) and biotic factors (such as species competition). Today very few biomes are unmodified by human activity, through agriculture, by mining, or pollution of the air. There are large variations in biomass productivity globally; tropical rain forests are very productive: 2200g/m^2/yr compared with semi desert: 90g/m^2/yr.

Change within ecosystems

The components of ecosystems are in a **state of balance** (**equilibrium**) with the physical environment. Ecosystems are dynamic and will evolve over time, as factors affecting them change. Ecosystems in Britain have changed since the last ice age in response to changes in climate.

The biosphere and ecosystems 2

A piece of bare land, perhaps exposed by deglaciation or a rock slide will over time become vegetated. The first plants, lichens, mosses, or where a soils exists, 'weeds', are known as the **pioneer species** or **community**, gradually these early plants are replaced by larger shrubby plants. Each stage is known as a **sere**. The final sere which is in equilibrium with the environment is known as the **climatic climax vegetation** or **climax vegetation**. The chain of seral stages that an area goes through to reach the climax vegetation is known as the **prisere**. Where human activities prevent the climax vegetation becoming established, perhaps because of grazing, afforestation, deforestation, monoculture or over cultivation, then the resulting vegetation is known as a **plagioclimax**. A **biotic climax** is when animals have prevented the climax vegetation from becoming established.

Four kinds of **primary succession** (where no vegetation was present before) have been recognised, (i) a **lithosere** (bare rock or rock slide), (ii) a **psammosere** (found on sand dunes), (iii) a **hydrosere**, found in ponds or lakes and (iv) a **halosere** in salt marshes. All these are frequently examined, and you will find questions on them in this unit. **Secondary succession** occurs where land has been colonised by plants, but is no longer managed by people and ultimately the climax vegetation may re-establish. This happens when areas are no longer subject to grazing.

Human activity is concerned with manipulating the environment by focusing production on those products that people find useful; food, timber or animals. The exploitation of the ecosystem will be limited by the existing land capacity and improvements that can be made to the soil. This may involve fundamental changes in the components present in the ecosystem, for example the changes that are made by subsistence rice farmers cultivating terraced slopes in the Philippines. The carrying capacity of land is higher where humans are largely primary consumers (vegetarians) as in parts of Asia, where they eat lots of rice, than where humans are tertiary consumers, which results in the inefficient conversion of plants to meat. Some consequences of human economic activity may be experienced a long way from the source area, such as atmospheric pollution in Scandinavia from the U.K.

It is important that you learn case study material from the biome type set in the syllabus and also local case studies of heath lands, uplands and marshes. Remember, examples that you have studied in the field or locally will often gain more credit than those gleaned from text books.

The properties of soil

Soil is an integral part of the ecosystem. A soil is defined as biologically and chemically modified parent material, and comprises inorganic and organic matter, water and air. The **texture** of a soil refers to its inorganic content, and is the percentage of sand, silt and clay it contains. The **structure** of a soil is the way the soil particles aggregate or group together; for example a soil may have a blocky, platey or crumb structure. The soil structure will influence the drainage of a soil and the ease with which plants grow.

Soil types and development

Soils form from weathered material (**regolith**) with the addition of further material – principally organic matter – and from which some soluble material is removed. Different soil types are recognised by the number and character of the **soil horizons** that are found in the **soil profile**. The character of these horizons is determined by the soil forming processes of weathering, biotic and chemical processes and soil water movements. The six factors which control soil formation are climate, parent material, topography, biota, time and man. Ultimately, climate tends to dominate the soil forming process, so each biome is characterised by a **zonal** soil (a soil determined by climate and vegetation type). Soils that are not climatically determined are **intrazonal** (determined by factors other than climate) such as the rendzinas which form in limestone areas, or **azonal** (immature soils which are not in equilibrium with their environment).

2 *The biosphere and ecosystems*

REVISION SUMMARY

Most questions refer to zonal soils such as **podsols, ferrallitic** (rainforest) or **ferruginous** (savanna) soils, so you must be aware of the processes (such as podsolisation and weathering) that lead to their formation in each biome and be able to draw their soil profiles. The diagram below shows the type of detail you ought to know and be able to sketch in an exam.

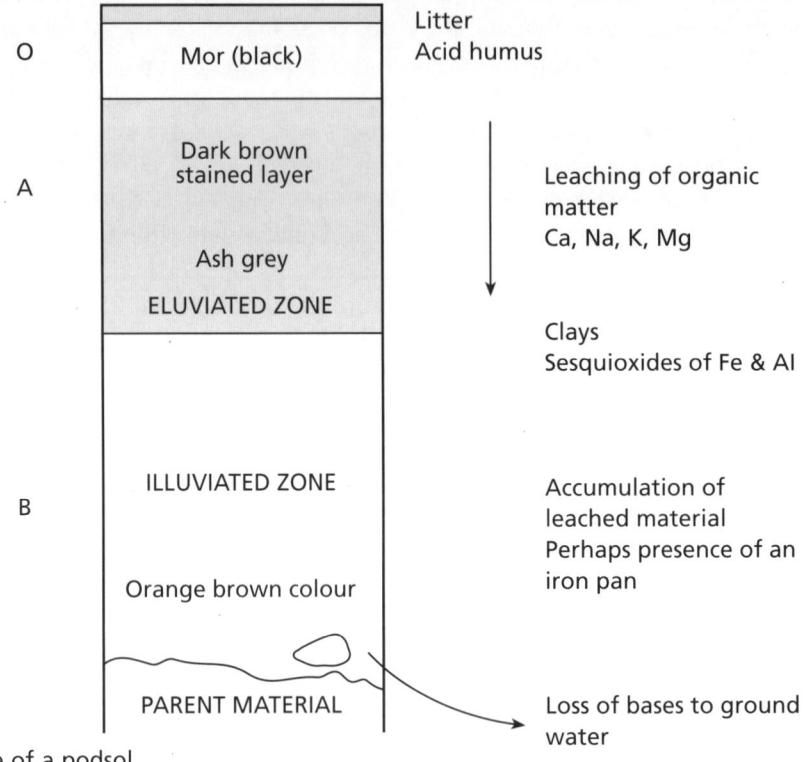

Soil profile of a podsol

Be aware of how the processes of weathering influence a soil's depth and characteristics. Particularly important is how climate influences the **water budget** which is the amount of water available to pass through the soil through the **precipitation to potential evaporation ratio** (**P:Pet**). A high P:Pet results in leaching and the **eluviation** ('e' for exit) of material from upper horizons; this material is then deposited in the a lower horizon (usually the 'B' horizon) by the process of **illuviation** ('i' for in). When Pet exceeds P, soluble material will be drawn towards the surface by capillary action, as for example, in the continental grasslands during summer.

Other important areas are the changes brought about by humans to soils by cultivation (interruption of the **organic cycle**, soil erosion changes in structure and many more), irrigation (**salinisation** may be an unwanted side effect), fertilisation, and soil acidification caused by increased industrial pollution. Many syllabuses also require an understanding of **soil catenas** (the sequences of soil types that are found on slopes resulting from processes of mass movement and water movements and conditions). You should ensure you know examples of these.

If you need to revise this subject more thoroughly, see the relevant topics in the *Letts* A level *Geography Study Guide*.

The biosphere and ecosystems

QUESTIONS

1 (a) Define **two** of the following terms and give an example of each: (6)

   zonal soil;

   intrazonal soil;

   azonal soil.

   (b) Describe the characteristics of **one named** zonal soil. (5)

   (c) Describe how the soil identified in (b) influences agricultural practices. (4)
   AEB

2 (a) Examine Fig. 1 (World Distribution of Northern Coniferous Forests (Taiga)).

   (i) Describe the geographical distribution of the Northern Coniferous Forests (Taiga) and explain why trees do not grow beyond the northern limit shown on the map. (3)

   (ii) With the aid of a sketch, describe the characteristics of coniferous trees and describe how they are adapted to their environment. (5)

   (b) With the aid of a diagram, describe the physical characteristics of podsols, the soils which have developed in these areas, and explain why they are difficult soils for farming. (7)

   (c) Explain the commercial importance of the Northern Coniferous Forests (Taiga) and describe methods taken to conserve timber stocks. (5)

Fig. 1

2 The biosphere and ecosystems

QUESTIONS

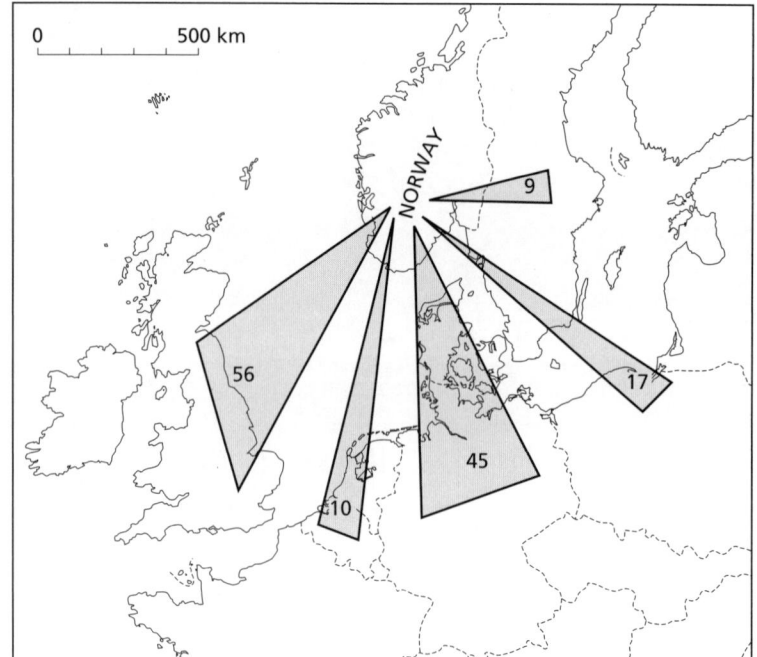

Fig. 2

(d) Examine Fig. 2, showing the main source areas of sulphur falling in acid rain over Norway. Explain the causes of this pollution and its significance to the forestry industry of Norway. (5)

SEB

3 (a) (i) Define the terms:

soil texture;

soil pH. (4)

(ii) What are the processes of illuviation and eluviation? (3)

(b) Using the data provided in Fig. 3 (overleaf), account for the differing relationships between vegetation, soils and biomass. (8)

(c) Critically examine the possible global effects of continued human interference within the Tropical rain forest ecosystem. (10)

UCLES

4 Figs. 4 and 5 (overleaf) represent diagrammatically the relative volume of nutrient flows (arrows) and the relative volume of stored nutrients (circles) in two major ecosystems.

(a) What do you understand by the term *biomass*? (1)

(b) Explain briefly the process of leaching. (2)

(c) Give one reason why the input of nutrients from weathered rock (d) is so much greater under Equatorial forest than under Temperate coniferous forest. (1)

(d) Explain why the litter store (L) is smaller under Equatorial Forest than under Temperate Coniferous Forest. (1)

The biosphere and ecosystems

QUESTIONS

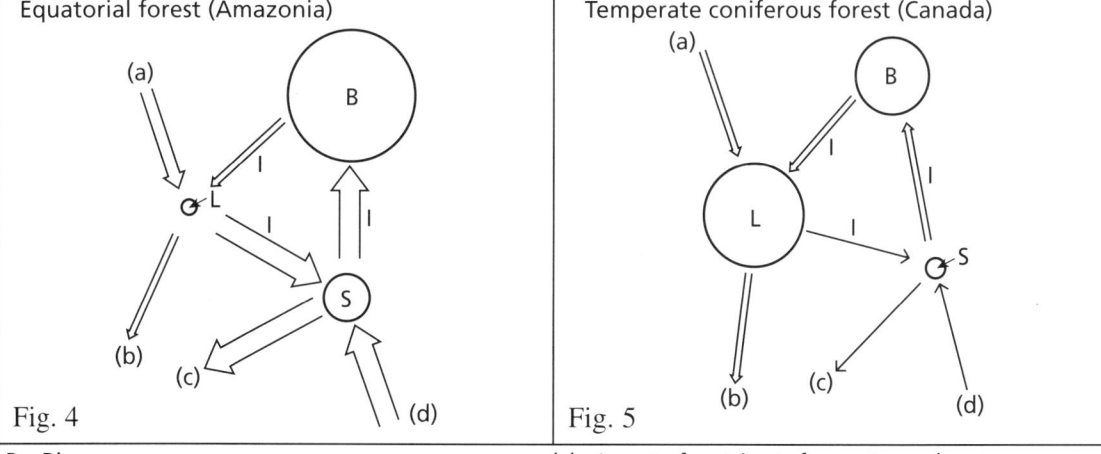

Vegetation	Tropical rain forest	Temperate deciduous woodland	Temperate coniferous woodland
Soil Profile	Litter — thin humus layer; Red colour with high iron and aluminium content coarse texture; Parent Material weathered to great depth	Litter — mull humus; Brown colour becoming lighter with depth; Some leaching; Parent Material	Pine needles — mor humus; Pale grey horizon with considerable leaching; Darker horizon with iron and humus enrichment; Parent Material
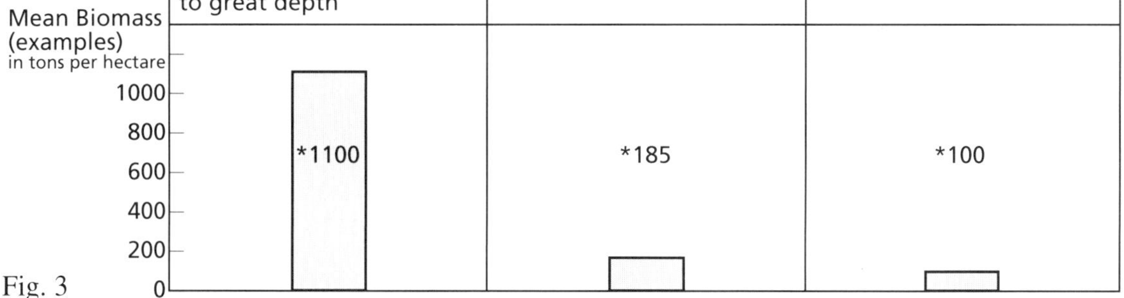 Mean Biomass (examples) in tons per hectare	*1100	*185	*100

*Actual figures

Fig. 3

Fig. 4 — Equatorial forest (Amazonia)

Fig. 5 — Temperate coniferous forest (Canada)

B: Biomass
L: Litter
S: Soil
I: Internal transport of nutrients

(a) Input of nutrients from atmosphere
(b) Loss of nutrients in run off
(c) Loss of nutrients by leaching
(d) Input of nutrients from weathered rock

The relative volume of nutrient flow and store is indicated by the width of arrows and size of circles respectively.

2 The biosphere and ecosystems

QUESTIONS

(e) Why is the loss of nutrients in runoff (b) so small under forest ecosystems? (1)

(f) Describe and explain two likely effects on the soil store (S) of wholesale clearance of Equatorial Forest for grazing purposes. (2)

Oxford & Cambridge (Specimen)

5 (a) Discuss the factors which encourage the development of an ecosystem to its climax stage. (12)

(b) Why might this climax stage not be achieved? (13)

ULEAC

6 (a) Illustrate how the flow of energy and circulation of nutrients vary between two contrasting ecosystems. (15)

(b) With reference to **one** of the following:

   a salt marsh,
   a dune system,
   a deciduous woodland,
   an area of moorland,

illustrate how people have affected the movement of energy and nutrients and with what consequences for the ecosystem. (10)

AEB

7 **Either**, (a) Compare the development of **two** contrasting seres. (20)

Or, (b) With reference to **one** biome, illustrate the ways in which its interrelationships may be changed by natural and human factors. (20)

WJEC

8 Consider the view that desertification is a process in which the actions of people are as important as the effects of climate. (25)

Oxford & Cambridge (Specimen)

Hydrology and drainage basins 3

Introduction
- This subject is seen by most examining boards as one of the core areas of physical geography. The scale of study is a *regional* one, contrasting with the study of the atmosphere or plate tectonics which operates at a global scale.
- In general, examiners are likely to wish you to adopt a systems approach.
- Remember you must not isolate your knowledge. Although rivers are an important influence on the quantity and type of load, weathering and mass movement will also have an impact on load and valley shape. Good candidates always make links with other areas of the subject.

The hydrological cycle
Water moves between areas of storage, such as the seas, oceans and ice caps, by the processes of evaporation, condensation, precipitation and runoff. The hydrological system is in equilibrium at the global scale as evaporation is compensated for by a return flow to the seas. Rivers form an important sub system within the hydrological system, but account for only 0.0001% of the global water balance.

Drainage basin systems
A **drainage basin** is the area of land drained by a river and its tributaries; it is delimited by high ground known as the **watershed**. A river can be considered as part of a drainage basin system, in which there are inputs (water and weathered material), throughputs (mass movement and fluvial erosion and transport) and outputs from the system (deposition). A drainage basin is an open system, as energy and mass enter and leave the boundaries of the system. For example, in the drainage basin system there are energy and mass inputs (precipitation) and outputs (evaporation and flow to the sea). The system is in dynamic equilibrium, and changes of input or output will lead to positive or negative feedback, so a period of drought will lead to a reduction of discharge and ground storage. This state of equilibrium is shown by the water balance.

Precipitation = Discharge + Evaporation +/- changes in storage

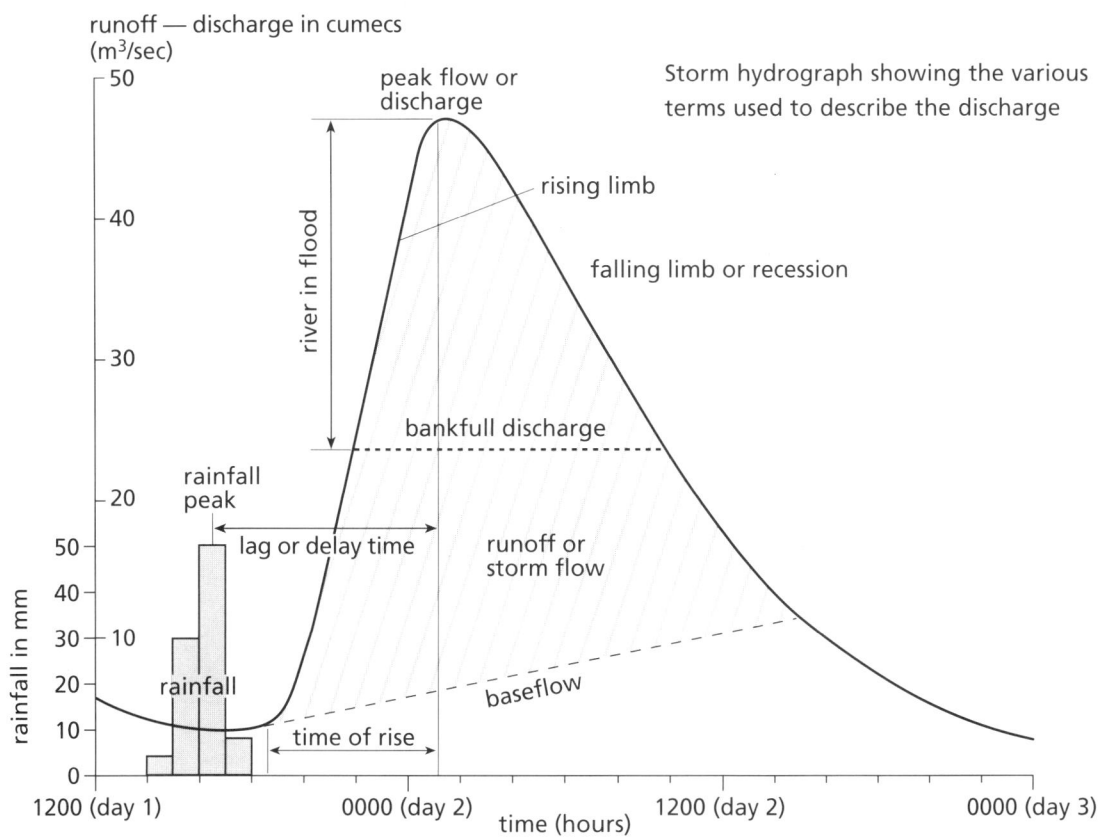

Storm hydrograph showing the various terms used to describe the discharge

3 Hydrology and drainage basins

REVISION SUMMARY

Discharge is the volume of water passing a point in a river channel normally measured in cubic metres (m^3) per second. River flow is determined by climate, geology, soil type and relief and may vary spatially and temporally. Rivers flowing over a permeable rock will have lower flows than similar rivers flowing over impermeable rocks. **Hydrographs** show the variation of discharge over time. Variations in flow over the course of a year reflect seasonal changes in runoff conditions, such as snow melt and summer drought, and are referred to as the river's **regime**. Hydrographs may also show changes in river flow over shorter periods of time, and a **flood hydrograph** shows how a river's discharge responds to a specific precipitation event. The appearance and response of the hydrograph will reflect the size and shape of the drainage basin, vegetation cover or land use type, geology, temperature, and relief.

You should be aware of how precipitation reaches the channel, and be able to define terms like **overland flow**, **through flow**, **base flow**, **infiltration rate**, **interception** and the temporary storage of water within the drainage basin system. Overland flow causes sheet and gully erosion; this is particularly marked on slopes that are cultivated or have been cleared by people, so overland flow is involved in slope evolution.

Drainage basin morphometry

Geographers traditionally used descriptive accounts of the river channel and its basin. **Drainage basin morphometry** is the quantitative description of drainage basins, and allows contrasts and similarities to be made more scientifically. These quantitative measures are a function and reflection of the work undertaken by the river. Horton undertook a study of the relation of **stream order** and drainage basin area, mean stream length and numbers of stream. The **bifurcation ratio** stems from stream order; it is the number of streams of one order divided by the next stream order. Most drainage basins have an order 2.5 – 3.5, but thin drainage basins will be dominated by the main stream. **Drainage density** is a measure of the length of streams in a drainage basin in relation to its area, and is determined by time, geology, land use and climate.

Stream flow and stream processes

Two types of river flow exist, **laminar** and **turbulent**. Natural streams are characterised by turbulent flow. Flow will be fastest away from the banks and beds of the river because of the effects of friction. It is important to be aware of the patterns of flow in varying channel shapes (straight, meandering, shallow and deep) as this influences the energy available to erode and transport the load of a river. **Channel geometry** can be described using a number of quantitative measures and these can be related to other variables such as velocity. A measure of channel efficiency is the **hydraulic radius** (this is the ratio of cross-sectional area to the **wetted perimeter** [the boundary between the river and its bed]). **Manning's roughness equation** relates channel geometry, and particularly the roughness of the bed, to velocity. Clearly, upland streams with their irregular channel and coarse bedload will function very differently from smooth artificial drainage channels. You should be able to describe the processes of erosion (**corrasion** [abrasion], **corrosion**, **hydraulic action**), transportation (the entrainment by rolling of the **traction** or **bedload**, **saltation**, **suspension** and **solution**) and deposition. The **Hjulstrom curve** shows the relationship between the velocity of a river and the ability to carry and erode material of different sizes (the **competent velocity** and **critical tractive force**). The important feature of Hjulstrom's curve is that the river is able to carry larger material as the velocity increases; however, much higher velocities are required to erode clay-sized material than to deposit it; this is because clays are cohesive and consequently difficult to entrain. There is a progressive reduction or **comminution** of load downstream due to attrition.

Erosion has three main components, **vertical down cutting**, **lateral erosion** and **headward erosion**, the relative dominance of each of these gives rise respectively to features such as gorges, meanders and spring sapping.

Hydrology and drainage basins 3

REVISION SUMMARY

Landforms in river valleys
A characteristic set of landforms results due to the interrelationships acting within the fluvial system. You need to be able to describe and discuss the formation of the following landform features. Make sure you can produce labelled diagrams of them. The following list includes most of the features you should know about: longitudinal and cross-sectional variations in valley shape; variations in channel plan (straight, braided, anastomosing, and meandering); the characteristic patterns of flow, climate, and load with which they are associated; and landforms such as waterfalls, rapids, gorges, interlocking spurs, floodplains, bluffs, riffles, point bars, oxbow lakes, levées, deltas. Many of these will be familiar from lower levels of study and details can be found in any good text or the *Letts A level Study Guide*.

Changes to the drainage basin over time
Fluvial systems are in equilibrium with their environment, with erosion being equal to the rate of deposition; a **graded profile** is one that allows this balance to be maintained. However, a graded profile does not have to have a smooth concave longitudinal profile.

Any alterations in the river's energy over time, such as **base level** changes due either to **eustatic** changes (sea level change caused by climatic change) or **isostatic** (tectonic in origin) or a change in discharge because of climatic change, will result in changes to the longitudinal profile and landforms present within the basin. **Rejuvenation**, resulting from a fall in base level, results in **incised meanders**, **terraces** and **river capture**.

Human influence on the hydrology, regime and processes of a drainage basin
The effect of human activity on drainage basins is a very important area of the subject. It is very important that you have case study material to back up your points, and that you do not compartmentalise your knowledge, but refer to the processes acting in the basin outlined above. Water may be abstracted from groundwater storage, or runoff may be stored in reservoirs; both will have consequences for the hydrology of rivers. Rivers are also important as a means of waste disposal, which is diametrically opposed to their use as a water supply. Interference in the drainage basin system may be designed to prevent floods, facilitate river transport (for example in Europe) or to provide water for irrigation or power schemes. In many cases there may be a conflict between the different users of the water, and this is most pronounced where water is in short supply such as in areas like Israel and Jordan.

Unintentional effects on the hydrology of drainage basins have resulted from changes in land use, such as urbanisation or agricultural use. These may increase discharge and sediment supply. Potential solutions to problems such as flooding must therefore consider drainage basins as a interrelated and integrated system.

If you need to revise this subject more thoroughly, see the relevant topics in the *Letts A level Geography Study Guide*.

3 Hydrology and drainage basins

QUESTIONS

1 (a) (i) With reference to river channels, define the terms

discharge;

hydraulic radius. (4)

(ii) Briefly describe the factors that affect a river's velocity. (3)

(b) Using Fig. 1 which shows a simple hydrograph:

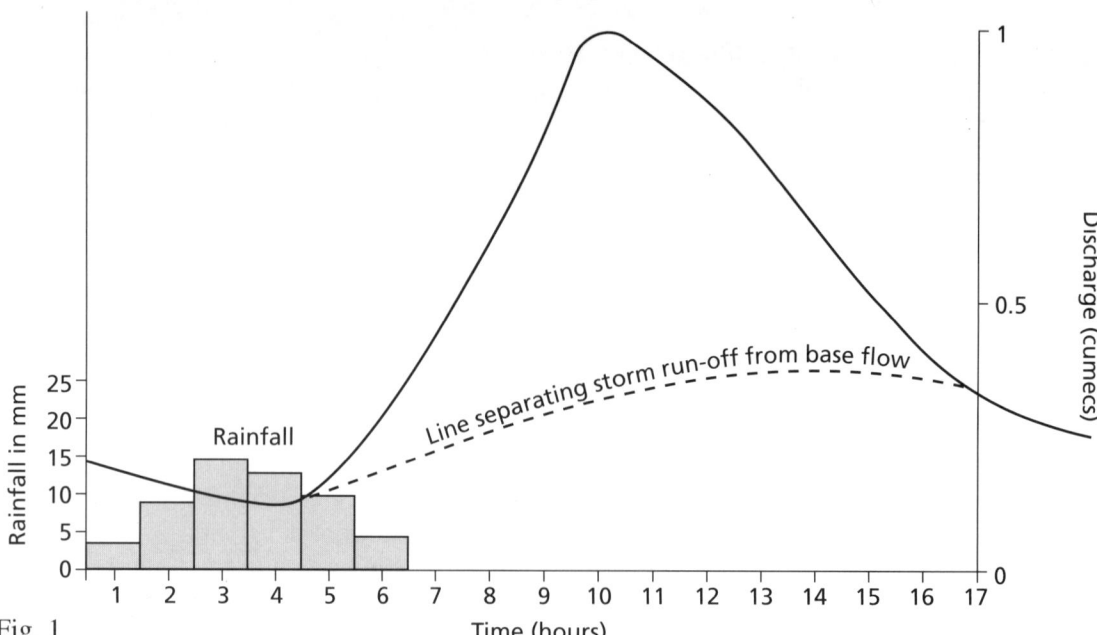

Fig. 1

(i) define what is meant by lag time and calculate it; (3)

(ii) suggest reasons for the relatively large contribution to the peak on the hydrograph of the storm run-off as compared with baseflow. (5)

(c) Under what conditions do rivers flood? How can a study of these conditions be used to predict and to limit the effects of floods? (10)

UCLES

2 (a) (i) Outline the factors that affect the infiltration capacity of soils. (4)

(ii) What is understood by the term soil moisture deficit? Under what conditions in Britain is there likely to be *no* soil moisture deficit? (3)

(b) Study photographs A and B (opposite).
Draw an idealised sketch cross-section of each river channel. Your diagrams should be annotated and indicate: channel shape, nature of the banks, type of river flow, channel landforms, any erosive or depositional process, nature of discharge. (8)

(c) Evaluate the relative importance of river processes and slope processes in influencing the shape of the cross-sections of river valleys. What other factors may be important? (10)

UCLES

Hydrology and drainage basins

QUESTIONS

Photograph A   *Acknowledgement:* R. J. Small

Photograph B   *Acknowledgement:* R. J. Small

3 Hydrology and drainage basins

QUESTIONS

3 Study Fig. 2, which shows river velocity at selected locations in a cross-section of a river channel.

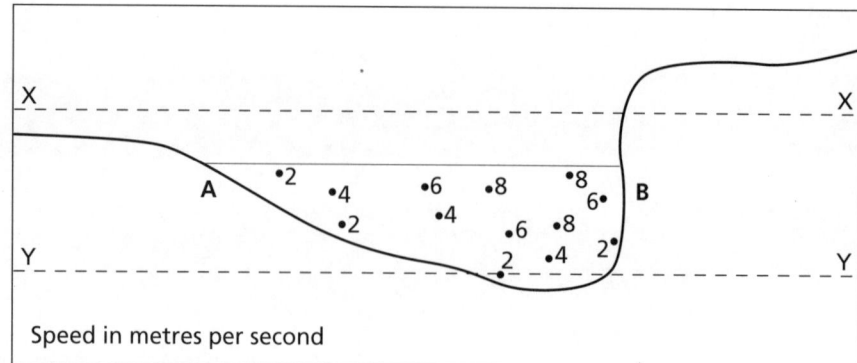

Fig. 2

(a) (i) Plot the isovels (lines of equal velocity) for 2, 4, 6, 8 metres per second. (4)

(ii) Explain the distribution of velocity shown in the channel cross-section. (3)

(b) With reference to Fig. 2, describe and explain the river processes at:

(i) A

(ii) B (6)

(c) What factors control:

(i) channel shape;

(ii) river velocity? (6)

(d) Describe and explain the changes likely to occur to the channel characteristics if the river level were to:

(i) rise to flood level XX;

(ii) fall to drought level YY. (6)

ULEAC

4 The reasons for the occurrence of floods and droughts often relate to both physical and human factors. Explain why this is so and describe ways in which such hazards may be prevented or at least ameliorated. (25)

Oxford & Cambridge (Specimen)

5 Examine the factors which affect the pattern of drainage in a catchment area. (25)

ULEAC

Arid and semi-arid environments 4

This is not a core topic, but appears on some syllabuses as an option. Brief revision notes are included to indicate the area of study. The questions and answer section will help focus your revision.

REVISION SUMMARY

The distribution of arid lands
Climatic reasons, for example high pressure cells, distance from oceans and rain shadow help determine the global locations of arid areas. During the **Pleistocene** arid areas were probably more restricted. The climate within arid areas varies tremendously, but all have precipitation levels below potential evapotranspiration (P<Pet) so they have a moisture deficit. Rainfall is often very variable.

Landscape forming processes
Weathering is very important in arid areas, the principal forms are **insolation weathering** (leading to **exfoliation** and **granular disintegration**) and **salt weathering**. Three other major processes operate in these landscapes – **slope evolution**, **aeolian** and **fluvial processes**. Wind moves sand by **suspension**, **saltation**, and **surface creep**. Coarse material cannot be moved by the wind so is left on a deflation surface. The lack of vegetation and moisture makes aeolian erosion and deposition a key process. The influence of geological structure on the landscape is very clear, for example in the Grand Canyon, Arizona, and the block faulted landscape of Nevada.

Landforms
You should be able to discuss the contribution of geomorphological processes to both sand deserts (**erg**) and rocky deserts (**reg** and **hamada**). Sand deserts make up only 12% of the land surface in arid areas. Reg consists of small sized rocky debris, whereas hamada have clear rock platforms and coarse sized debris. The appearance and formation of the following features must be known, **inselbergs**, **mesas** and **buttes**. Aeolian deposition produces features which vary greatly in size from sand ripples to **barchan** and **seif** dunes. Dune form is a function of wind direction and sand supply. Common features produced by aeolian erosion are **yardangs** and pedestal rocks (**zeugen**). Features resulting largely from fluvial processes are **wadis**, **arroyos**, **alluvial fans**, **playa lakes** and **bahadas**. Rivers in arid areas are classified as **exogenous**, originating outside the area (e.g. The Nile), **endoreic** where the drainage ends in an evaporating saline inland lake, such as the Dead Sea, and **ephemeral** which flow only after rainfall and then dry up.

Climatic change
The climate has not been constant over the last 100 000 years, and during the Pleistocene it was wetter. During **pluvials** lakes were larger, e.g. Great Salt Lake is a remnant of a much larger lake, Lake Bonneville. Some features of arid areas must have formed under wetter climates than are now found, for example alluvial fans.

Human activity within desert ecosystems
Arid areas are fragile and the climate leads to distinctive vegetation and soil types. Questions often focus on **desertification** and its possible causes (over grazing or climatic change), and evidence for these changes. Irrigation of arid areas coupled with increased population pressure has resulted in soil erosion and **salinisation** of soils.

If you need to revise this subject more thoroughly, see the relevant topics in the *Letts* A level *Geography Study Guide*.

4 Arid and semi-arid environments

QUESTIONS

1. (a) Rivers in deserts may be classified as *exogenous*, *ephemeral*, *endorheic*. Define any **two** of these terms. (4)

   (b) (i) Describe the landforms which result from the irregular and intense precipitation experienced in desert areas. (4)

   (ii) Account for the formation of one of these landforms. (4)

   (c) In what ways might irregular and intense precipitation be used by desert peoples? (3)
   AEB

2. Refer to the photograph.

   (a) Examine the contributions of lithospheric, atmospheric and hydrological processes to the formation of:

   (i) the dunes at X;

   (ii) the desert plain at Y;

   (iii) the slope and its features at Z.

   (b) How may human activity influence the movement of a dune such as that at X? (25)
   Oxford & Cambridge (Specimen)

Arid and semi-arid environments 4

QUESTIONS

3 (a) Outline the processes which are responsible for the formation of the type of desert landscape shown in Fig. 1. (9)

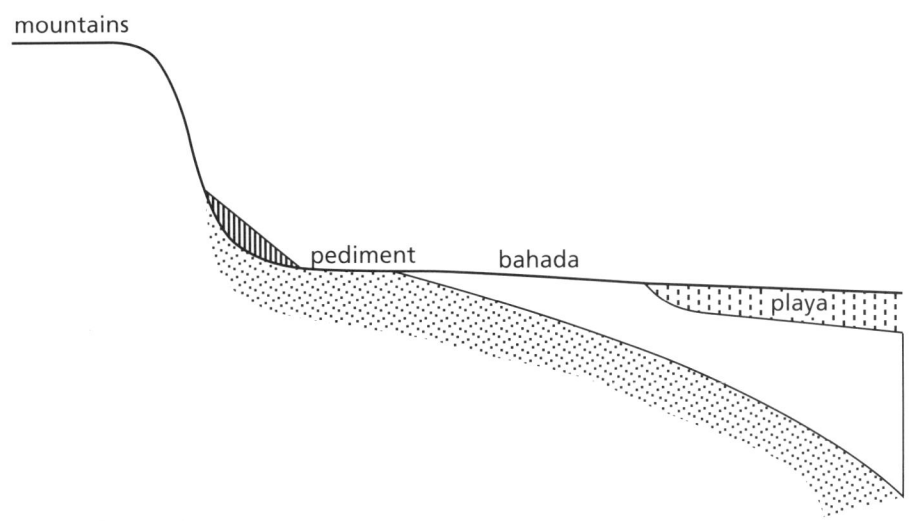

Fig. 1

    (b) Discuss the contention that hot desert landforms do not indicate a simple relationship with present day climate. (16)

UCLES

4 (a) Describe and explain the major characteristics of the soils of arid and semi-arid areas of the world. (15)

    (b) (i) Define the term *soil salinisation*. (2)

       (ii) Explain why salinisation often accompanies the irrigation of arid and semi-arid regions of the world. (8)

Oxford

5 (a) In what forms and under what conditions does running water occur in hot deserts in present times? (9)

    (b) Using examples, discuss the extent to which existing desert landforms are the product of present as against past erosional and depositional activity. (16)

UCLES

5 Coastal environments

REVISION SUMMARY

Marine processes

Coastal landforms are a result of complex and interacting processes such as marine, **subaerial weathering**, **tectonic**, **sedimentary**, **biotic** and **geological** processes, modified by human activity. Energy inputs into the coastal system come from waves, tides and currents. Wind blowing over the sea generates waves. Wave characteristics are determined by the **fetch**, and the strength and duration of the wind. In shallow water the orbiting motion of water within the wave is prevented, and as a result, the height of the wave increases, and eventually breaks, releasing its energy. Two main kinds of wave are recognised; (i) the **plunging** or destructive wave where the **backwash** is stronger than the **swash** and (ii) the **spilling** or constructive wave where the reverse is true. **Wave crests** have a directional component and this may lead to a movement of material along the coast by **longshore drift** (LSD) where a dominant wind exists. When the wave enters shallower water, the forward velocity is reduced. If part of the wave crest is slowed earlier (for example approaching shallow water in front of a headland) the wave crest will be distorted and will curve round the headland. This is known as **wave refraction**. Wave refraction concentrates erosion on the headland, wave crests travelling into the bay will have a lower height and the energy will be spread out over a greater area. Mid latitudes are more influenced by storm waves than low latitudes. Tides are oscillations of the sea surface and cause breaking waves to reach different parts of the beach profile. Areas are characterised as being **macrotidal** (much of the British Isles) **mesotidal**, or **microtidal**.

Waves erode by **abrasion** (corrasion), **quarrying** (where the hydraulic power of the wave loosens and removes blocks from cliffs), and **corrosion** (or solution). The latter is important on limestone, where it may lead to features as **lapiés** (pinnacles), and pronounced notches. **Attrition** wears material into finer and more rounded particles. Organic processes are important in stabilising **dunes**, **salt marshes** and **mangrove swamps**. In tropical areas coral flourishes in warm, shallow waters and leads to the construction of **reefs** and **atolls**.

Erosional landforms

The form of cliffs varies greatly according to whether erosion is active or whether mass movement and subaerial processes dominate. The shape is also influenced greatly by **lithology** and **geological structure** (see the diagram). Lithological differences are readily exploited by the sea, giving a finely etched cliff profile.

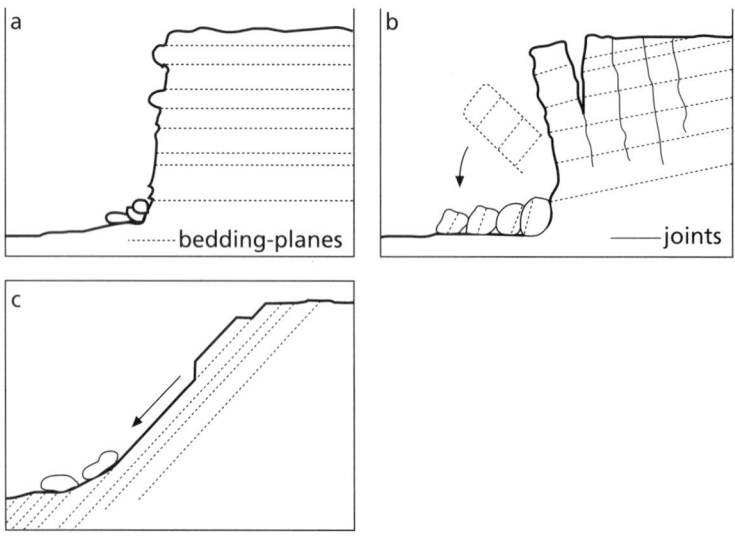

The influence of geological structure on cliff profiles

Coastal environments 5

The retreat of cliffs leaves a **wave cut platform**, a rocky, eroded shelf which may be covered in rock debris. The impact of differential erosion is clear along coastlines at a variety of scales. Where faults or other weaknesses occur in headlands, caves, arches and stacks may form, and on a larger scale Lulworth Cove is an example of lithological control on a **concordant** coast (geological structure runs parallel to the coast).

Depositional landforms

Beach sediment comes from various sources. There are differences in the sediment supply in tropical and temperate areas. In the former the dominance of chemical weathering results in fine clay-sized material, whereas in higher latitudes there is shingle and sand. A systems approach has been applied to sediment on coastlines. A **littoral cell** (a segment of shoreline) can be regarded as having a **sediment budget**, with both natural and human inputs (such as beach nourishment) and outputs (storm removal or LSD) of sediment.

Beaches are an important depositional feature, and show a variety of profiles and forms depending on the processes operating. For example, steep beaches may be produced in areas with a high percentage of shingle; this is because the porosity of the sediment reduces the effectiveness of backwash. In plan, beaches show **swash aligned** (concave) or **drift** features. Erosion by plunging waves produces erosional terraces and **offshore bars**, whereas building gives berms. In sandy macrotidal areas, ridges and runnels may result. Beaches can have seasonal variations in their profile as the wave type may vary between seasons. Another feature showing the complex nature of deposition is the barrier beach, which forms in microtidal areas as an offshore barrier island but is driven landward across a bay. Features resulting from LSD are **bars**, **spits** and **tombolos**. In association with spits and estuaries are **salt marshes**, and behind sandy beaches, **sand dunes** may form from wind blown material. Salt marshes result from the **flocculation** of suspended clays as they enter saline water, combined with the contribution of the specialised halophytic vegetation in binding the unconsolidated sediment.

Isostatic and eustatic movements

Features result from changes in sea level (**eustatic**) movements or crustal (**isostatic**) movements; in the UK, this is usually caused by deglaciation. In Scotland, isostatic recovery has produced raised beaches, whereas in the south the relative rise in sea level has flooded valleys forming rias, and **fjards** (lowland glaciated valleys). **Fjords** are found in uplands where glaciated valleys are flooded. On a larger scale the interaction of tectonic folding, plus sea level rise has formed **Dalmatian coastlines**, e.g. former Yugoslavia where tectonic folds run parallel to the shore, and **Atlantic coastlines** in southern Ireland where a discordant trend of folds has produced a coastline of bays and headlands.

The human influence on coastlines

Erosion of our coastline is an emotional issue, and local politicians, influenced by their electorate, spend large sums of money protecting agricultural, tourist and residential interests from disappearing into the sea, even when the economic benefits are far from clear. Methods adopted to protect shorelines include **sea walls**, **groynes**, **breakwaters**, **revetments** and **beach nourishment** (adding sand). Development of coastal areas will have an impact on coastlines, but so do changes in the sediment supply by damming, for example the Akosombo dam on the River Volta. Protecting one segment of the coastline may cause profound changes to another area: this can be seen where the construction of groynes disrupts the sediment supply further down the coast, and without the protection of the beach, erosion may result.

Other subjects that may be examined are: managing the coastline for recreation, reclamation of land and pollution control.

REVISION SUMMARY

If you need to revise this subject more thoroughly, see the relevant topics in the *Letts A level Geography Study Guide*.

5 Coastal environments

QUESTIONS

1 Fig. 1. shows an area of coast in the south of England.

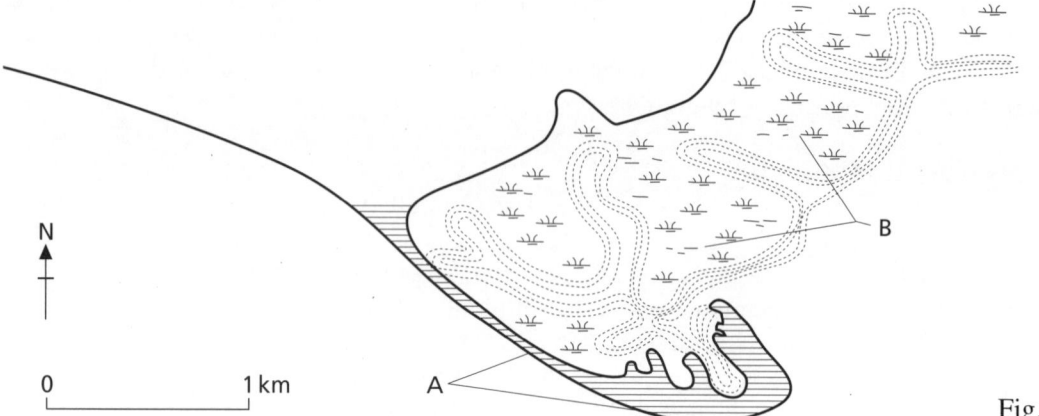

Fig. 1

(a) With reference to the shaded feature labelled A:

　(i) name the landform feature; (2)

　(ii) explain the origin and form of the feature. (8)

(b) With reference to the areas labelled B:

　(i) describe the form and character of the areas; (3)

　(ii) describe the vegetation that would be characteristic of the areas; (5)

　(iii) describe and explain how the areas will gradually evolve over time. (7)

Oxford

2 With the help of information provided on the 1: 50 000 O.S. Map Extract 792/159 on page 71 and the accompanying geological map, describe and attempt explanations of the origin of the coastal landforms between The Knave (431863) and Oxwich Point (513851). (25)

Oxford & Cambridge

3 (a) Define the following terms in the context of processes of coastal erosion:

　(i) corrasion (abrasion);

　(ii) corrosion;

　(iii) hydraulic action. (5)

(b) Describe features of cliff coastlines that are created mainly by processes of coastal erosion and suggest how such features develop. (13)

(c) With reference to one or more named area(s), show how coastal depositional processes may change the shape of a coastline. (7)

NEAB

Coastal environments 5

QUESTIONS

4 (a) Explain how the longshore movement of beach material may affect the shape of the coastline. (10)

(b) In what ways do people modify this process and with what effects? (15)
AEB

5 'The morphology of the coast is the product of rock type and structure rather than of process.' Discuss. (25)
ULEAC

6 (a) Explain the processes by which waves erode cliffed coastlines. (9)

(b) Identify and explain those features of coastal landforms in areas you have studied that can be attributed to changes in sea level. Estimate the effects upon coastal landforms of a rise in sea level that might occur through the effects of global warming. (16)
UCLES

6 Glacial and periglacial environments

REVISION SUMMARY

This is not a core topic, but appears on some syllabuses as an option. Brief revision notes are included to indicate the area of study. The questions and answer section will help focus your revision.

The current and past distribution of periglacial and glaciated landscapes

During the **Quaternary**, glaciated areas were much more extensive. The 'Ice Age' did not consist of a single cold period, but was characterised by warmer **interglacials**. The glacial budget refers to the balance between inputs of snow and outputs by melting or evaporation; when accumulation exceeds **ablation** (loss of ice) the glacier will advance. The 'Ice Age' has left a clear legacy within all areas of the British Isles, not only of upland features like corries, but also modified drainage patterns and till deposits. Periglacial areas were also more extensive in the past and many areas within Europe show relic features such as **coombe rock** and valleys within limestone areas as evidence.

Landscape forming processes

You should be aware of the significance of weathering in both these areas. In periglacial areas processes of mass movement such as **gelifluction** (or **solifluction**) and **sheet wash** determine much of slope evolution. The absence of vegetation means that fine material may be removed by aeolian processes. Glaciers erode by abrasion and plucking, though these processes are much more complex than was first thought; the importance of intense weathering such as **nivation** and **dilatation** must be discussed in this context. Glacial meltwater plays a very important role, not only in creating fluvioglacial deposits (such as **kames**, **eskers** and **sandur**), but also in allowing **regelation**, freeze thaw and chemical solution to occur.

Landforms

Clearly, features of both upland and lowland glaciation should be known. In upland areas the obvious erosional features are **corries** (cirques or cwms), glaciated troughs, **roches moutonnées** and **arêtes**. In lowland areas there are erosional features such as **knock and lochan** landscapes, **rock drumlins**, as well as depositional features like **till**, **moraines** and **drumlins**. Glaciation has had a major impact on the development and **modification of drainage**; a good case study is drainage in the North Yorkshire Moors where there were glacial spillways and ice margin lakes.

Key features within periglacial areas are those resulting from **cryoturbation** (involutions, ice wedges), and where ice becomes segregated into lenses within the ground, features like **patterned ground** and **pingoes** result. Students should be aware of the hydrology of rivers within tundra areas. Here there are huge variations in seasonal discharge and load resulting from thawing in late spring and early summer and the channel geometry has to accommodate this. Streams are often braided and have steep gradients.

Human activity within tundra ecosystems

Arctic and Antarctic areas are fragile and the climate leads to distinctive vegetations and soil types. Human exploitation of these landscapes is significant, as glaciated landscapes are important for leisure activities and power generation. Tundra areas have seen significant mineral exploitation, for example oil in Alaska and coal mining in Spitzbergen, though the latter also had political and strategic implications. There are large technical problems to overcome in these areas because of the **permafrost**. Buildings must be raised on stilts or have refrigeration units installed in their foundations to prevent the ground from melting, pipelines must be lagged to prevent freezing of water supplies or sewage, and communications can be difficult due to blizzards. As tundra vegetation is extremely fragile and slow growing, it can take years for vegetation to re-establish after being disrupted. The recognition of the delicate balance of these ecosystems has led to careful monitoring of development, and the prohibition of mineral exploitation in the Antarctic.

> If you need to revise this subject more thoroughly, see the relevant topics in the *Letts A level Geography Study Guide*.

Glacial and periglacial environments 6

1 (a) Describe the climate of Tundra regions. (4)

   (b) Describe the following Tundra landforms and explain how periglacial processes have produced them:

   (i) patterned ground;

   (ii) pingoes. (12)

   (c) Give an explanatory account of the vegetation succession which may have occurred in lowland Tundra regions as a result of changing climatic conditions during deglaciation. Suggest how the succession may have been influenced by local conditions such as water-filled depressions and bare rock surfaces. (9)

   NEAB

2 Refer to the O.S. 1 : 50 000 Map Extract No. 531/41, Glencoe on page 70. What evidence can you deduce from the map that this area has been glaciated? (25)

   Oxford & Cambridge

3 Refer to the accompanying aerial photograph which shows a variety of upland landforms in the Cairngorm region of Scotland.

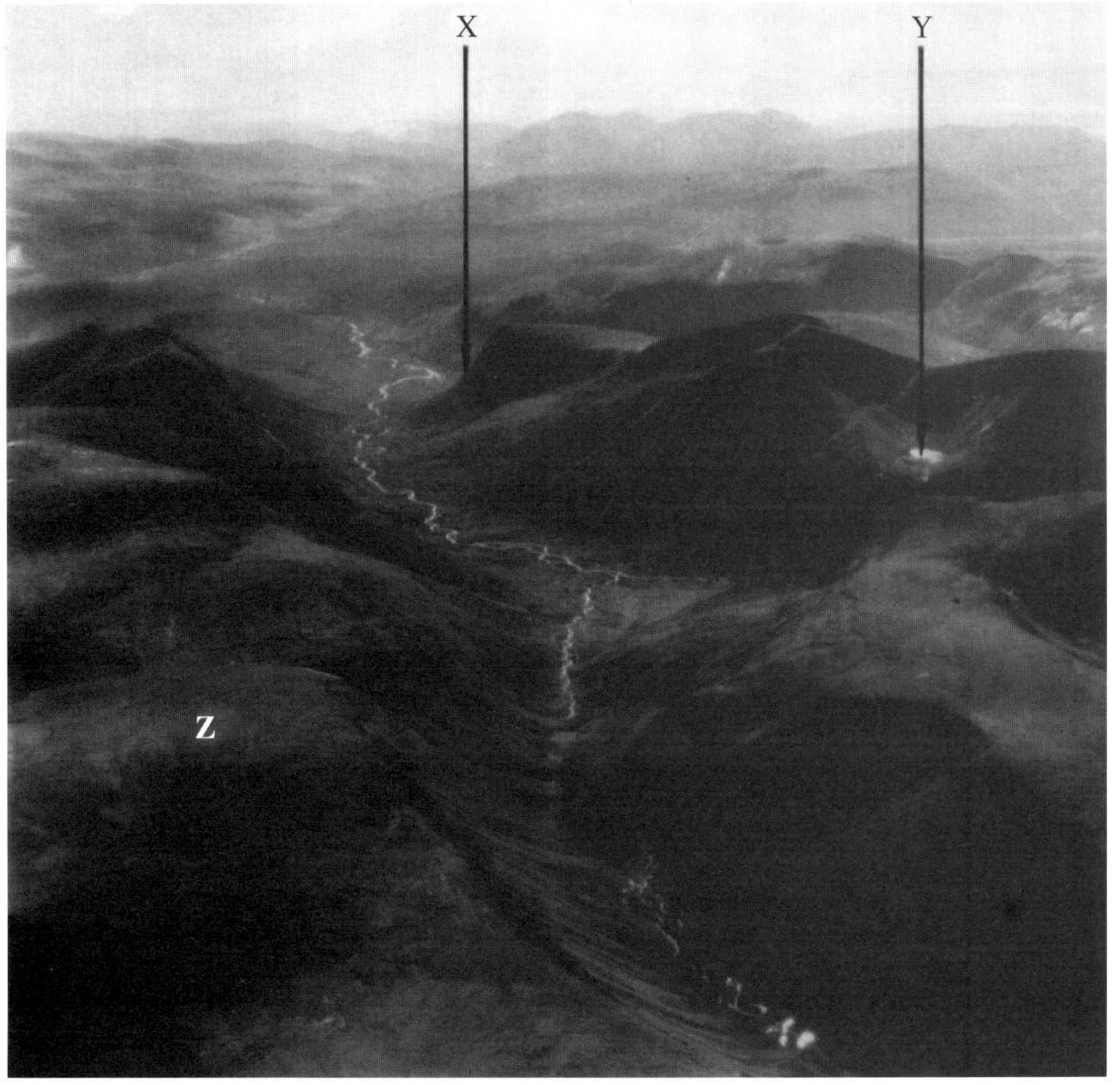

Cambridge University Collection of Air Photographs: copyright reserved

6 Glacial and periglacial environments

QUESTIONS

Attempt explanations of the following features:

(i) the steep slope at X;

(ii) the hollow at Y;

(iii) the surfaces at Z. (25)

Oxford & Cambridge

4 Study Fig. 1 which classifies the processes of glacier movement.

Fig. 1

```
                    Glacier movement
                    ┌──────┴──────┐
              Warm glaciers   Cold glaciers
                    ↓              ↓
              Basal movement   Internal flow
         ┌────────┼────────┐
    Regulation  Creep  Extending &
                       compressing flow
                       ↓
                    Surges ←
```

(a) Explain:

(i) the basis for distinguishing between 'warm' and 'cold' glaciers; (2)

(ii) how areas of warm and cold glaciers differ in terms of their landforms. (4)

(b) By means of two annotated diagrams, show the variations in velocity likely to be encountered across and within a 'warm' glacier. (4)

(c) Distinguish between the **three** processes of 'basal movement'. (6)

(d) Define the term 'surges' and identify the circumstances which can give rise to them. (4)

WJEC

(NB: this is a Board-written question in response to a centre-devised module.)

5 (a) Describe the landforms produced by glacial meltwater and by periglacial streams. (9)

(b) In **either** glacial **or** periglacial environments, to what extent do geomorphological processes and landforms both aid and hinder the human exploitation of these areas? (16)

UCLES

6 (a) Describe and explain the characteristic features of the following fluvioglacial landforms:

(i) eskers; (4)

(ii) kames; (4)

(iii) outwash plains. (4)

(b) Explain how to distinguish between glacial deposits, fluvioglacial deposits and soliflucted head deposits. (9)

(c) Explain why many fluvioglacial deposits pose problems for commercial exploitation. (4)

Oxford

The lithosphere and plate tectonics 7

REVISION SUMMARY

The structure of the earth

The earth is composed of concentric shells of rock-like materials. The composition and layered structure is deduced from seismic waves. At the centre is a solid inner **core** made of iron and nickel, the outer core is liquid. The next shell is the **mantle**, which is able to deform plastically. The outermost layer is the **crust**. There are two types of crust: **oceanic crust** (less than 10 km thick and with a basaltic composition), and **continental crust** (average thickness 33 km, though it can reach 65 km). Continental crust is less dense than oceanic crust.

Tectonic activity

'Tecton' is the Greek word for builder, so, plate tectonics concerns the construction of the earth's major landforms resulting from the movement of **lithospheric plates** across the earth's surface. The lower mantle is heated by radioactive decay which causes convection currents within the mantle. The up-welling of hot mantle rock leads to bulging and rifting of the crust, and the creation of fissure volcanoes, which are found along spreading ridges. The creation of crust at **constructive margins** is counter-balanced by the destruction of crustal rocks at **destructive margins**, where material is returned to the mantle by **subduction**. This boundary is marked by an oceanic trench, intrusion of batholiths and island arc volcanoes. Continental collisions, for example the Eurasian and African plates, create fold mountains as neither plate can sink. At conservative (passive boundaries) there is no crustal creation or destruction, but they are not uneventful. On these boundaries the crust is slipping past each other, but jerkily, due to friction. You will need to give detailed explanation of the processes (including folding and faulting) that occur at plate boundaries, as well as knowing the characteristic rock types and features found there. The evidence for the movement of plates includes; continental fit (the giant jigsaw of Pangaea), reconstruction of past climates, topographical features (such as spreading ridges and volcanoes), variations in heat flow and gravity, the distribution of earthquakes and volcanoes, and the symmetrical pattern of ages of the rocks on either side of mid-ocean ridges (shown by radiometric dating and palaeomagentism).

Natural hazards and human activity

Earthquakes are found at all types of plate boundary. Earthquake energy is measured on the **Richter scale**. The amount of damage an earthquake produces depends on a number of factors including, the depth of the **focus**, the proximity of the **epicentre** to highly populated areas, the geology and the likelihood of sediment liquefaction. The 1995 Kobe earthquake showed the vulnerability of supposedly earthquake-proof construction, in a developed country. Most people regard the forecasting of earthquakes as a remote possibility at present, so efforts must be made to ameliorate the effects on people when they occur. Volcanic activity cannot yet be reliably predicted, but increased monitoring of seismic activity, ground deformation and SO_2 emissions, has allowed vulcanologists to improve their predictions. The area surrounding Mt Pinatubo in the Philippines, was successfully evacuated in 1991. People have also had some success in diverting lava flows away from villages. Students should be aware of different kinds of volcanic eruption (effusive and explosive) as well as variations in volcanic products (for example gases, ashes, lavas, and pyroclastics) and the separate hazards they present.

Structure, process and landforms

Candidates should know about the influence of lithology and the basic geological structures (for example faulting, folding, inverted relief, grabens) and have case study material to illustrate them.

If you need to revise this subject more thoroughly, see the relevant topics in the Letts A level Geography Study Guide.

7 The lithosphere and plate tectonics

QUESTIONS

1. Study Fig. 1 which shows a constructive plate margin.

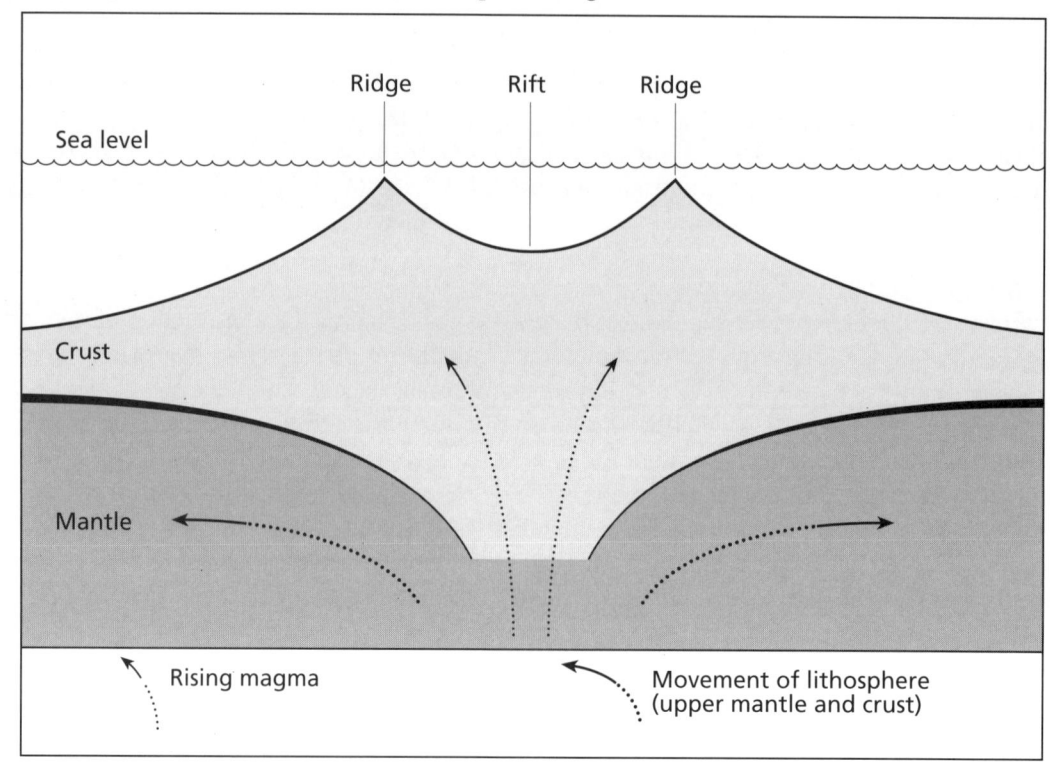

Fig. 1

(a) (i) Why is this type of margin called 'constructive'? (2)

   (ii) What causes the plates to move apart? (3)

   (iii) What evidence can be used to support the concept of sea-floor spreading? (3)

(b) With the aid of a diagram, describe the formation of the surface and subsurface features occurring at a destructive plate margin. (8)

(c) Why are subduction zones likely to experience periodic earthquakes of varying intensity? (5)

(d) Account for the difference in volcanic activity occurring at constructive and destructive plate margins. (4)

ULEAC

2. Critically examine the view that the presence of different types of plate margin may be recognised by the occurrence of distinctive landforms. (20)

WJEC

3. (a) Discuss the relationships between the form and composition of different types of volcano. (12)

   (b) Explain the global distribution of the different types of volcano. (8)

   (c) Discuss the hazards associated with different types of volcano. (5)

Oxford

The lithosphere and plate tectonics

QUESTIONS

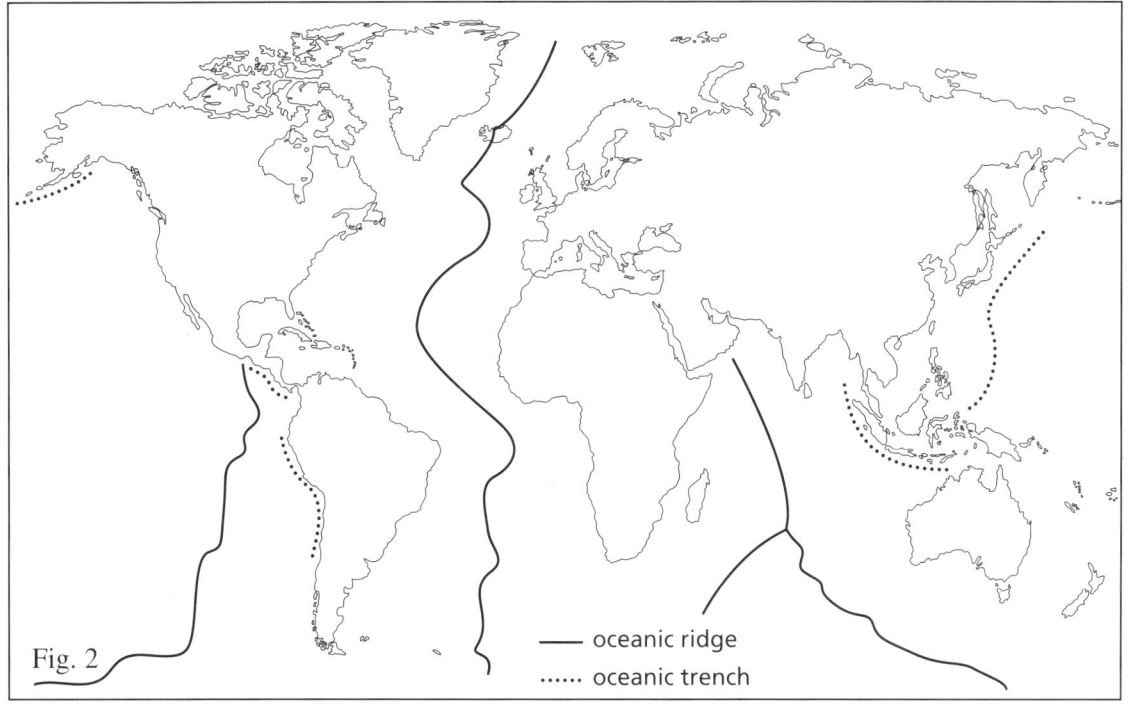

4 Study the world map (Fig. 2) below showing the distribution of the main oceanic ridges and trenches.

Fig. 2
— oceanic ridge
...... oceanic trench

(a) State what determines the position of these ridges and trenches. (2)

(b) (i) Name the process giving rise to ocean ridges. (1)

   (ii) By means of an annotated sketch explain their origin. (4)

   (iii) Explain why rifts are commonly associated with these ridges. (3)

(c) Explain why oceanic trenches are most often paralleled by island arcs. You may draw an annotated diagram to help your answer. (4)

(d) Suggest and briefly justify **three** different criteria that might be used in drawing up a classification of **one** natural hazard associated with these ridges and trenches. (6)

WJEC

5 (a) Study Figs. 3A and B (overleaf) showing the distribution of earthquakes and volcanoes near the Tonga islands.

   (i) Describe and account for the distribution of earthquakes and volcanoes shown. (15)

   (ii) People on Tonga are at risk from infrequent tsunamis (often incorrectly known as tidal waves). What are the physical and human factors which can contribute to this risk? (6)

(b) With the aid of a diagram, explain how island chains can result from 'hot spot' activity. (14)

NICCEA

7 The lithosphere and plate tectonics

QUESTIONS

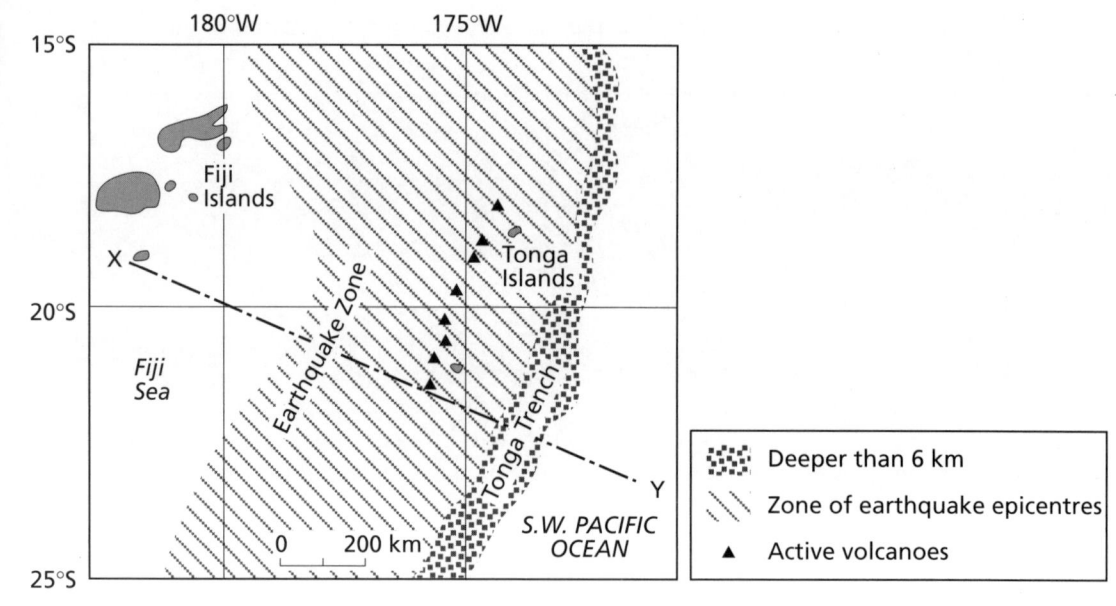

Fig. 3A *The distribution of earthquakes and active volcanoes between the Tonga Trench and Fiji Islands*

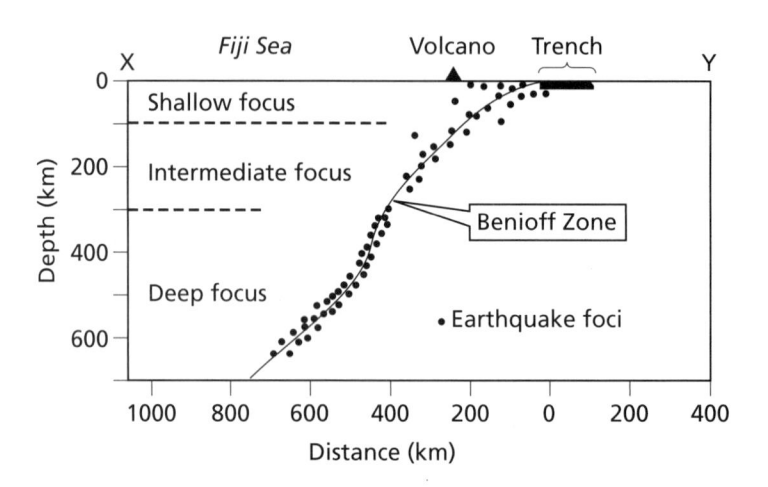

Fig. 3B *Cross section of earthquake foci along XY*

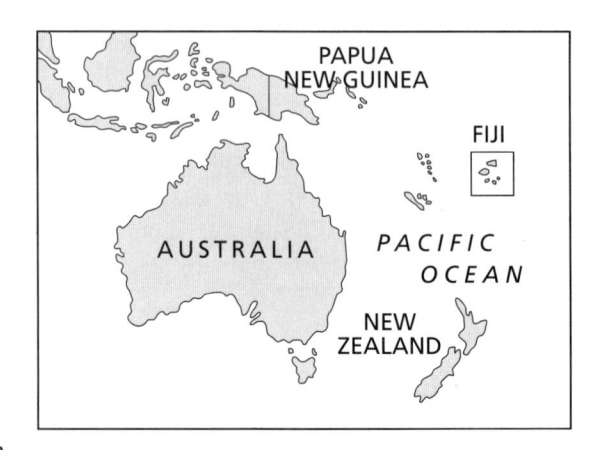

Fig. 3C *Location map*

Settlement 8

REVISION SUMMARY

You must know the difference between rural (hamlets and villages) and urban (towns and above) settlements. Criteria often used to describe 'urban' include: minimum populations, administrative areas designated by Government, percentage of population engaged in non-agricultural activities and urban functions and services present.

Rural

Although only 23% of the UK lives in a rural setting (1981), over 59% (1985) of the world's population lives in the countryside. The balance is tilted by countries in Asia such as China and India where about 72% of the population live in the country.

You need to know: why villages locate where they do. You must consider the **site** (actual space occupied) and **situation** (the surrounding area) and that both physical (e.g. water availability) and human (e.g. transport route) factors will influence this. You must study the three main **shapes** or **morphologies** rural settlements take such as **nucleated** or focused on one central point, (e.g. Co. Durham), **linear** or ribbon (elongated along a road), or **dispersed** or scattered (e.g. farming communities in North Wales). You must then look at what function rural settlements perform and how market towns serve and interact with their hinterlands: these can be explained by **Christaller's Central Place Theory** (C.P.T.). This helps explain the hierarchy of settlements and their spheres of influence.

It is then important to look at changes which occur and what their impact is, alongside the policies that have been used to shape development of settlements:

- in the Developed World. This includes rural service decline (e.g. shops and transport) in-migration from cities of commuters and second home ownership (suburbanisation of the countryside)
- in the Developing World. In particular, the continuing rural-urban migration and its effects ('brain' and 'brawn' drain).

Urban

Urban places are becoming globally more numerous although some Developed World countries are experiencing **counter-urbanisation** (i.e. dispersal of people from big cities to smaller settlements). There is a continuing shift in the balance of **'Millionaire'** cities to the Developing World; 61% of these are now found south of 40° N. This city growth is fuelled by rural-urban migration *and* natural increase (i.e. urbanisation and urban growth).

You need to know about the link between urbanisation and industrialisation which occurred in the Developed World in the nineteenth century and is now happening in the Developing World, as well as the distribution and density of populations within cities, and why the density decay curves vary between Developing and Developed World cities. These reflect differences in the economy, and wealth of citizens (Developed World households can afford more space) and their ability to afford transport (low in the Developing World).

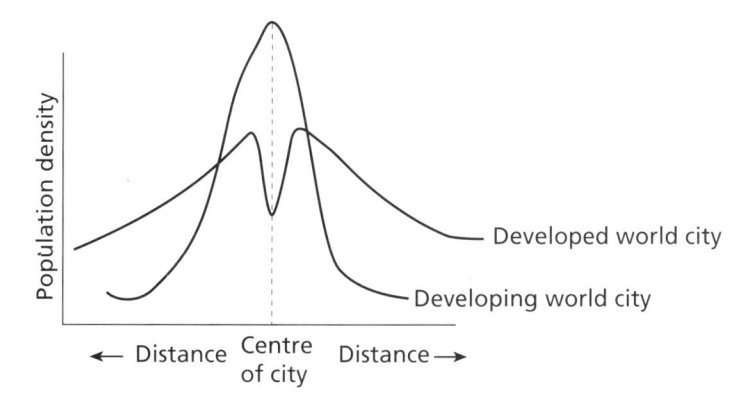

A comparison of population densities between Developed and Developing World cities

8 Settlement

REVISION SUMMARY

The location of industry within cities should be cross-referenced with the Industry section, concentrating on its drift from old nineteenth century sites to peripheral locations. You must also consider the location of **tertiary activities** (offices and shops) within a city: the intra-urban pattern. This will include the suburban **shopping hierarchy** and recent out-of-town (off centre) developments such as retail parks.

Models and theories help explain land use pattern. For example **Bid Rent Theory**, is the idea that different land uses compete for space, and that those which are in the best position to bid, i.e. those with most money, will get their desired location. Other models describe Developed World cities including: **Burgess** (concentric rings) which is the basis for the others, **Hoyt** (sectors), **Mann** (UK-wind direction) and **Harris-Ullman** (multiple nuclei). Burgess' model is shown below.

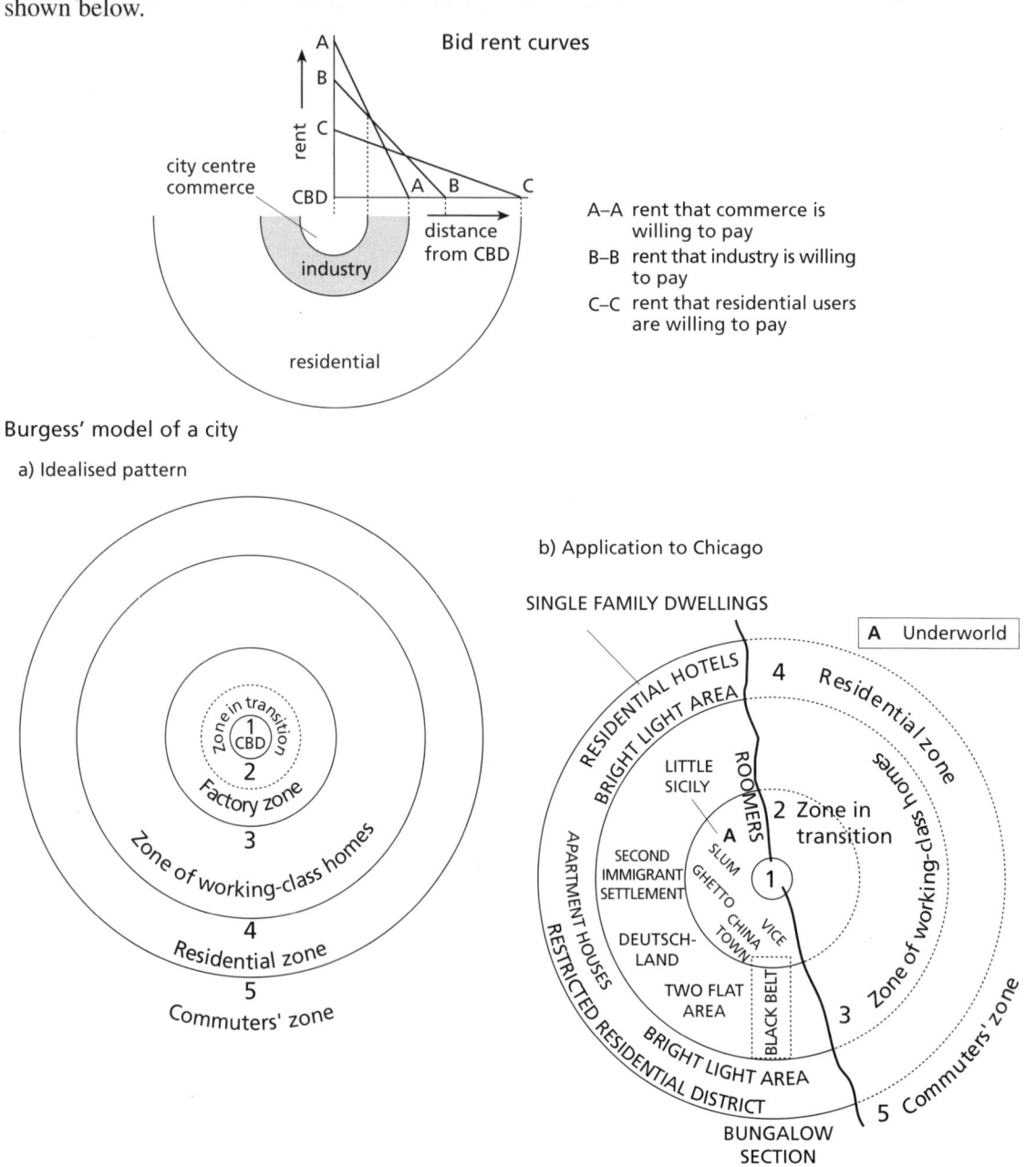

Bid rent curves

A–A rent that commerce is willing to pay
B–B rent that industry is willing to pay
C–C rent that residential users are willing to pay

Burgess' model of a city

a) Idealised pattern
1 CBD
2 Factory zone / Zone in transition
3 Zone of working-class homes
4 Residential zone
5 Commuters' zone

b) Application to Chicago

SINGLE FAMILY DWELLINGS
RESIDENTIAL HOTELS
BRIGHT LIGHT AREA
LITTLE SICILY
ROOMERS
SECOND IMMIGRANT SETTLEMENT
SLUM
GHETTO
CHINA TOWN
VICE
DEUTSCH-LAND
TWO FLAT AREA
BLACK BELT
APARTMENT HOUSES
RESTRICTED RESIDENTIAL DISTRICT
BRIGHT LIGHT AREA
BUNGALOW SECTION
A Underworld
2 Zone in transition
3 Zone of working-class homes
4 Residential zone
5 Commuters' zone

The basic conclusions from these are that different land uses concentrate in certain areas due to the attraction (similar) and repulsion (different) of other land uses, and competition between different land uses. Therefore, different socio-economic and ethnic areas develop, e.g. CBD, industrial, high, medium and low income residential.

Settlement 8

REVISION SUMMARY

There are also models relating to Developing World cities, for example the generalised model of a Latin American city (below), or McGee's model of a South Asian city.

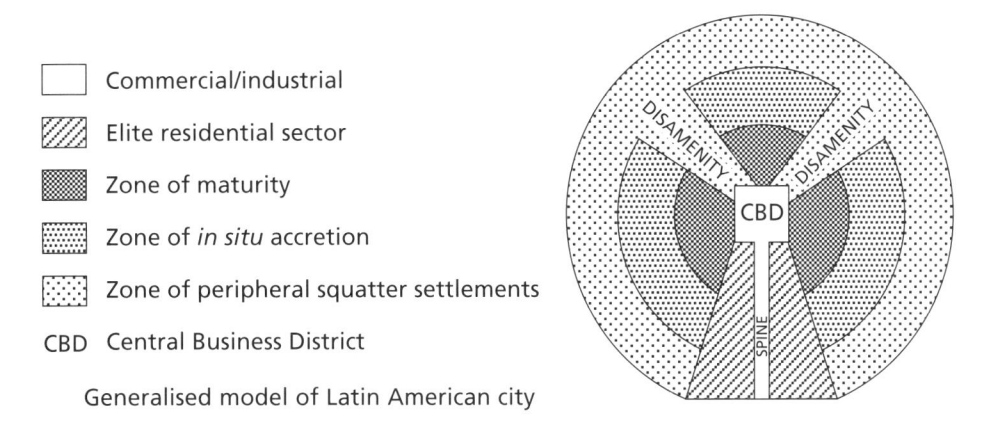

- Commercial/industrial
- Elite residential sector
- Zone of maturity
- Zone of *in situ* accretion
- Zone of peripheral squatter settlements
- CBD Central Business District

Generalised model of Latin American city

Some syllabuses, such as Oxford and Cambridge, require a similar study of Socialist Cities, but this is less common.

Developed World city problems and their possible solutions need to be studied. These include:

- counter-urbanisation, making cities redundant as people and activity disperse
- congestion and urban transport, including commuting
- inner city problems including unemployment, housing, crime, ethnic segregation and multiple deprivation
- conflict at the rural-urban fringe (the Green Belt) and suburbanisation of the countryside
- gentrification and city policy, including the inner city.

Also, Developing World city problems and possible solutions should be learnt. Often **Rank Size Rule** and **primacy** (primate cities) and the dominance of metropolitan concentrations are applied here, although they are just as relevant to Developed World cities. You must bear in mind:

- causes of rapid growth in primate cities (migration and natural increase)
- unemployment and underemployment
- provision of services such as housing, transport, sanitation and power
- pollution and waste disposal.

You need good detailed case studies to illustrate all these areas and concepts.

If you need to revise this subject more thoroughly, see the relevant topics in the *Letts A level Geography Study Guide*.

8 Settlement

QUESTIONS

1   (a) Study Fig. 1.

   (i) Describe and account for the pattern of land uses shown in the model.

   (ii) To what extent is the model an accurate representation of any city in the Developing World which you have studied? (10)

  (b) Fig. 2 shows the typical population structure of a city in the Developing World. Describe and account for this structure. (5)

  (c) Referring to a named city in the Developing World:

   (i) describe the environmental, social and economic problems it is currently facing;

   (ii) suggest ways in which these problems are being tackled. (10)

   SEB

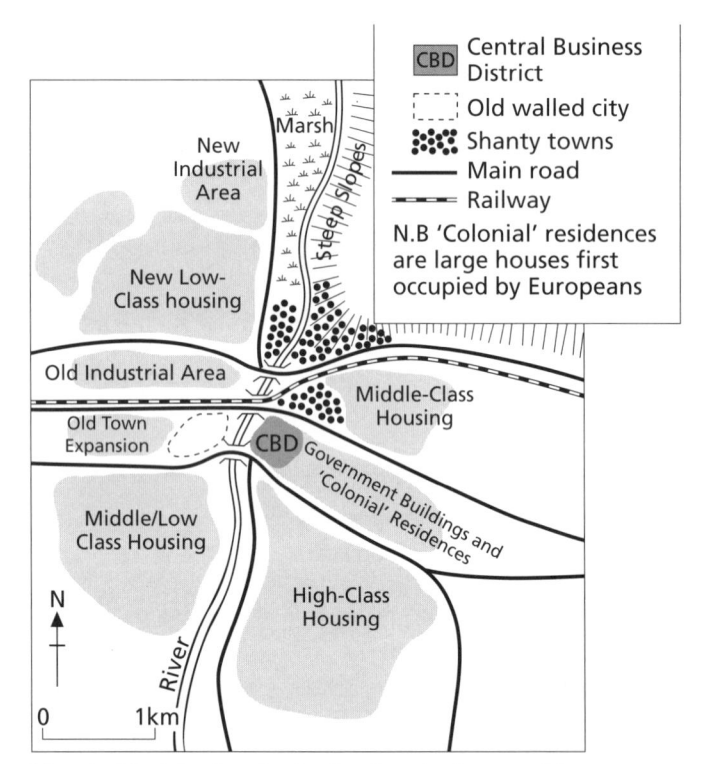

Fig. 1 *Model of a city in the developing world*

Fig. 2 *Age pyramid of a city in the developing world*

2   Fig. 3 opposite shows the changes that a county council is considering for an area of rural settlement in Eastern England.

  (a) Summarise the changes that are proposed for the location of

   (i) medical services; (2)

   (ii) shops. (2)

  (b) Why would a council decide to make the changes shown in the figure? Suggest **two** reasons. (4)

Settlement 8

Fig. 3 *County council proposed changes for an area of rural settlement in Eastern England*

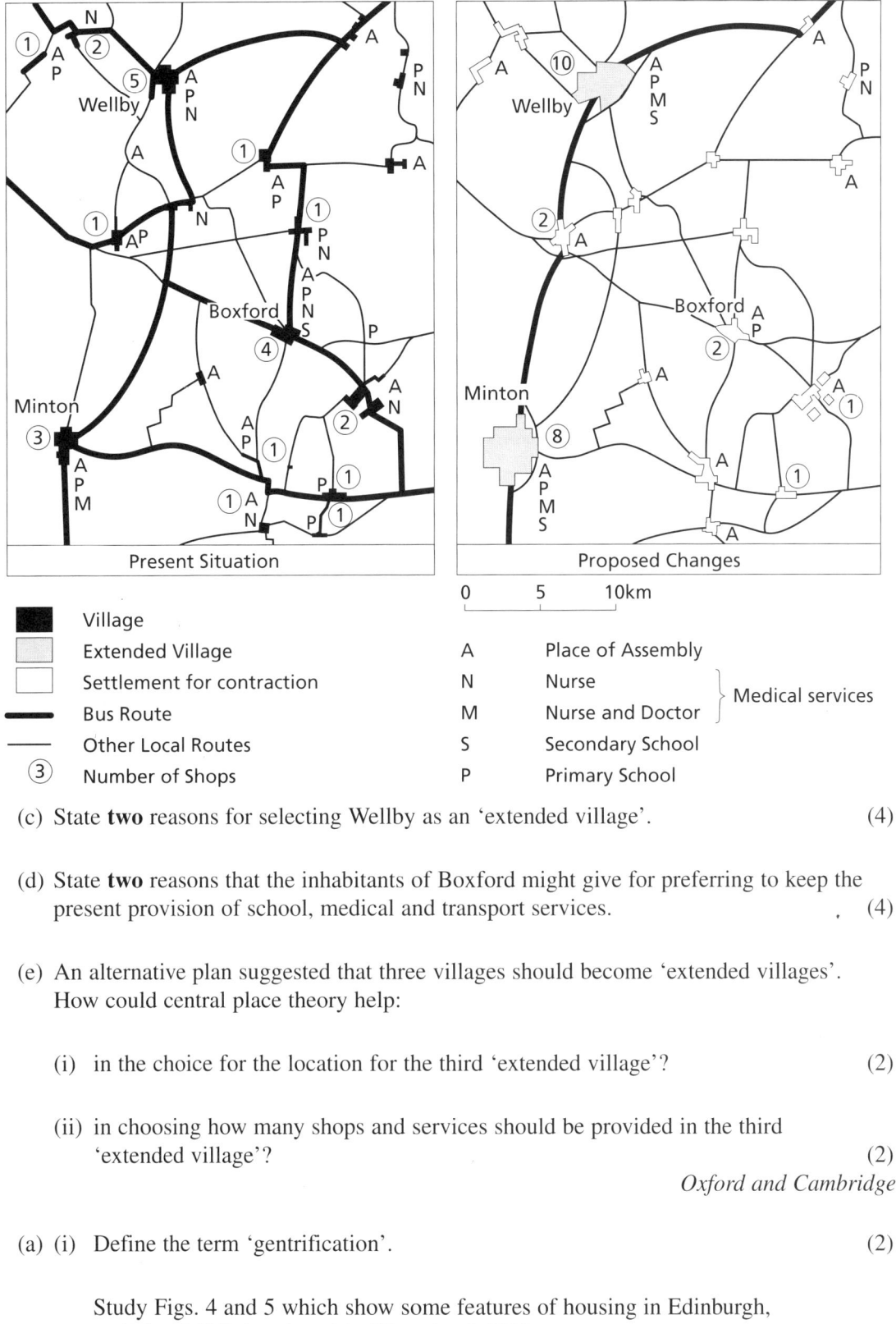

Present Situation | **Proposed Changes**

0   5   10km

- ■ Village
- ▨ Extended Village
- ☐ Settlement for contraction
- ━ Bus Route
- — Other Local Routes
- ③ Number of Shops

A   Place of Assembly
N   Nurse ⎫
M   Nurse and Doctor ⎬ Medical services
S   Secondary School
P   Primary School

(c) State **two** reasons for selecting Wellby as an 'extended village'. (4)

(d) State **two** reasons that the inhabitants of Boxford might give for preferring to keep the present provision of school, medical and transport services. (4)

(e) An alternative plan suggested that three villages should become 'extended villages'. How could central place theory help:

   (i) in the choice for the location for the third 'extended village'? (2)

   (ii) in choosing how many shops and services should be provided in the third 'extended village'? (2)

   Oxford and Cambridge

3 (a) (i) Define the term 'gentrification'. (2)

   Study Figs. 4 and 5 which show some features of housing in Edinburgh, and Fig. 6, Edinburgh and its Hinterland, 1898.

   (ii) Using the information provided in these figures, suggest one ward in Edinburgh where gentrification is likely to have occurred, and justify your suggestion. (2)

   (iii) Why does the process of gentrification take place in some Western cities? (3)

8 Settlement

QUESTIONS

(b) Refer to Figs. 4, 5 and 6.

(i) Outline the features of the distribution patterns shown on Figs. 4 and 5.

(ii) Suggest reasons that might help to explain the pattern shown on Fig. 4.

(iii) Suggest reasons for the levels of overcrowding mapped on Fig. 5 in Craigmillar (23) and Central Leith (19). (8)

1 St. Giles
2 Holyrood
3 George Square
4 Newington
5 Liberton
6 Morningside
7 Merchiston
8 Colinton
9 Sighthill
10 Gorgie - Dairy
11 Corstorphine
12 Murrayfield - Cramond
13 Pilton
14 St. Bernards
15 St. Andrews
16 Broughton
17 Calton
18 West Leith
19 Central Leith
20 South Leith
21 Craigentinny
22 Portobello
23 Craigmillar

% households sharing or lacking W.C.
- 14.1 - 19.1
- 9.6 - 14.0
- 5.0 - 9.5
- 0.2 - 4.9

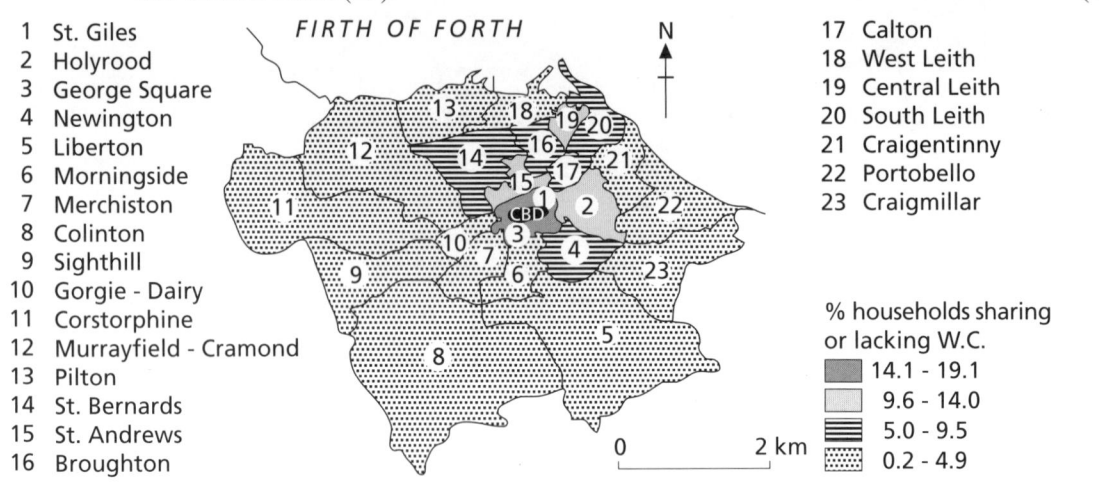

Fig. 4 *Distribution of households sharing or lacking W.C. facilities in Edinburgh, 1978*

% overcrowded households
- 10.1 - 13.2
- 7.1 - 10.0
- 4.1 - 7.0
- 1.1 - 4.0

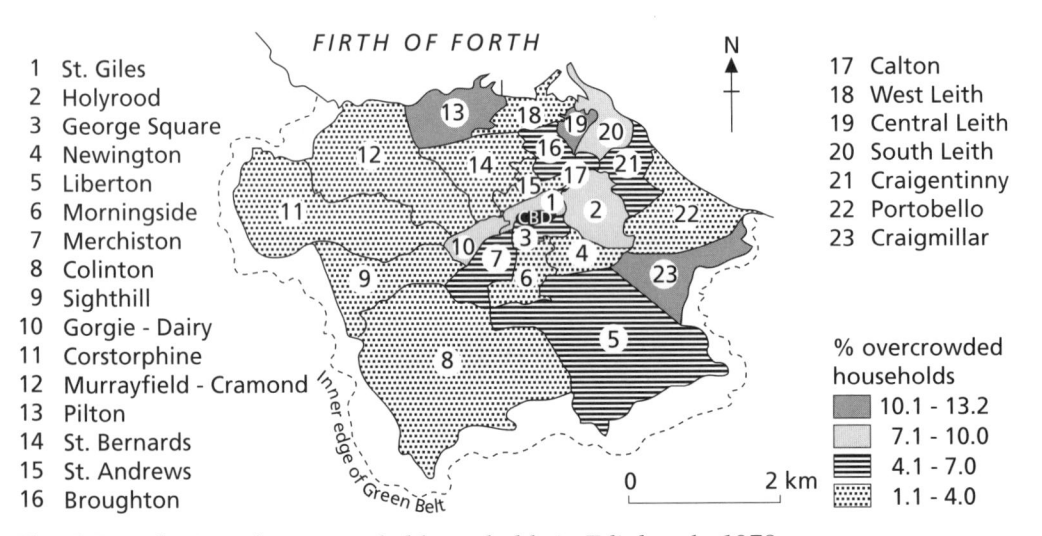

Fig. 5 *Distribution of overcrowded households in Edinburgh, 1978*

1 C.B.D.
2 Holyrood Park
3 Gorgie
4 New Town
5 Old Town
6 Morningside

Fig. 6 *Edinburgh and its hinterland, 1898*

Settlement 8

(c) By detailed reference to one or more examples, consider the extent to which the suburbanisation of the countryside has had positive social and environmental consequences. (10)

UCLES

4 Why is it difficult to counteract the tendency for areas of multiple deprivation to concentrate on inner city areas? (25)

Oxford and Cambridge

5 (a) With reference to Fig. 7 describe:

(i) the changes in the total population living in urban areas between 1965 and 1988;

(ii) the variations in the rate of growth of the urban population in the period 1965 to 1988. (6)

(b) What does Fig. 8 suggest about the relationship between urbanisation and economic development in 1988? Suggest reasons for your answer. (9)

(c) With reference to Fig. 9 and using your own knowledge, comment on and suggest reasons for:

(i) the number of cities showing population increase or decline between 1975-80 and 1985-90;

(ii) the distribution of cities showing population increase or decline between 1975-80 and 1985-90. (10)

NEAB

Fig. 7 *EEC countries: degree of urbanisation and growth, 1965–88*

8 Settlement

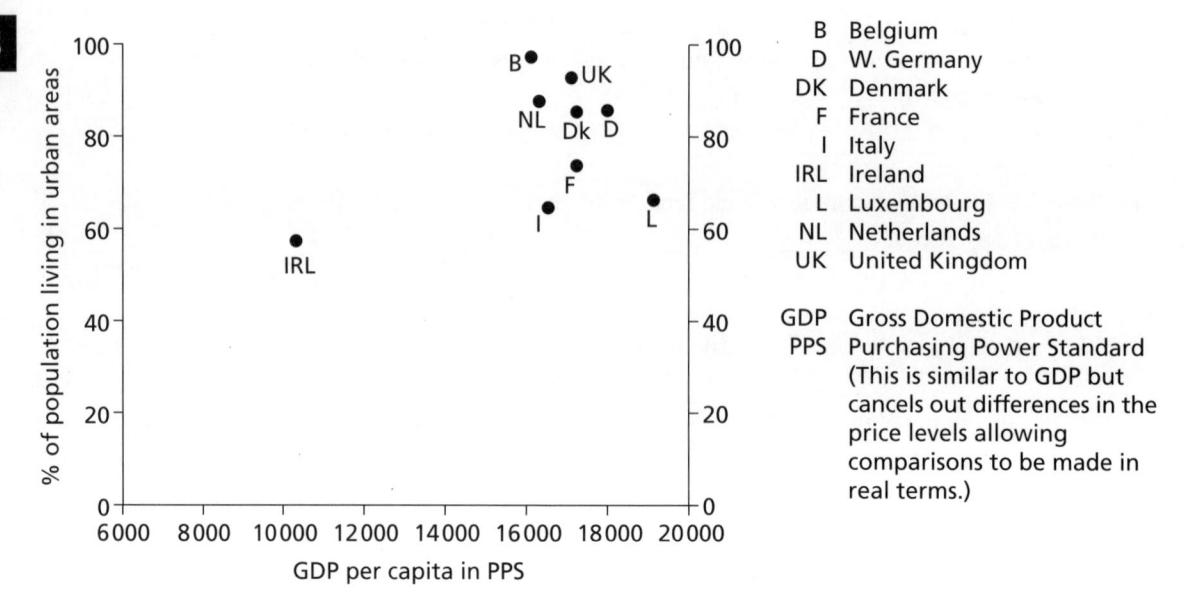

QUESTIONS

Fig. 8 *EEC countries: urbanisation and economic development, 1988*

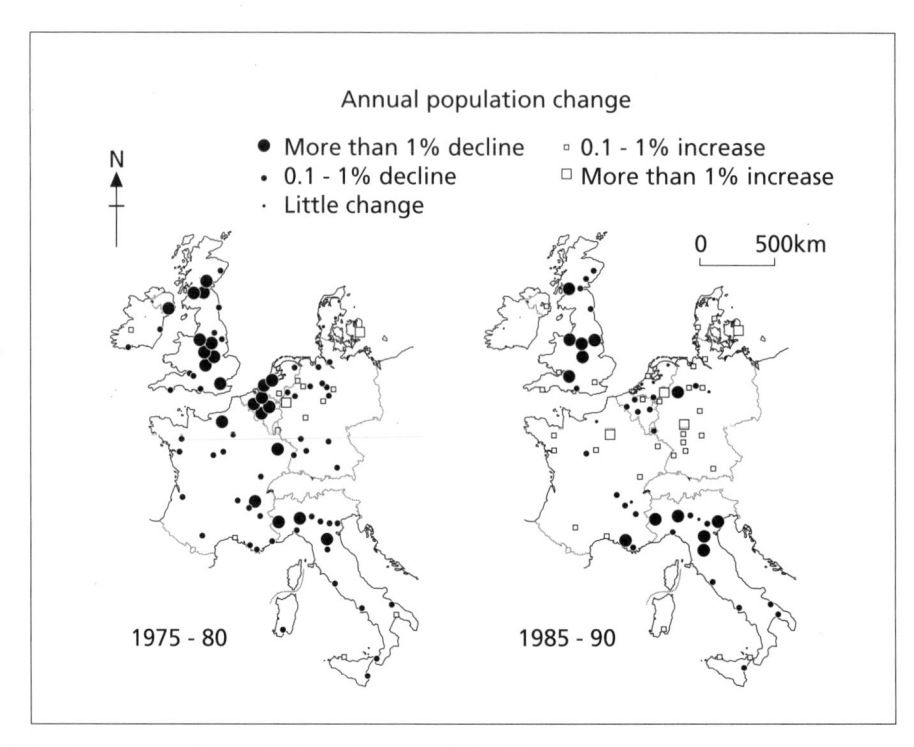

Fig. 9 *EEC cities: annual population change, 1975–90*

Population and resources 9

REVISION SUMMARY

Population is often in the news as it is an ongoing current affairs topic. With an ever increasing global media (e.g. CNN) giving us pictures of famine and disease, we question their causes. A simple answer, of course, is to blame the increase in population in the Developing World. Nearer to home, the question of how we can support an ageing society is becoming an important issue.

You need to know definitions such as **overpopulation**, which is often ill-defined yet is vital in understanding **carrying capacity** or **optimum population**. You must be careful in making the distinction between population **distribution** (where people are) and **density** (how many people there are in a given area).

The **Demographic Transition Model** (DTM) suggests how the population of a country develops through time, linking population change with economic development and the process of modernisation. This four-stage model is based on Western European countries' experience, and therefore has limitations when applied to developing countries. This is very evident when applying the model to Muslim countries, especially oil-rich ones. You must be able to evaluate the DTM not only in biological terms (i.e. the relationship between population and resources), but also through sociological understanding.

Age-Sex Population Pyramids are 'snapshots' of a population showing the make-up or structure of the population at that time. You can infer future changes to the population by looking at its present characteristics (e.g. if large numbers of young women are present [a high **reproductive capacity**], it is likely that birth rates will rise in the future). Connections with the DTM can also be made.

Population theories

Malthusian ideas: In 1798 Malthus published his *'Essay on the Principle of Population'*. This stressed that populations naturally outstrip their resources, as population grows at a geometric rate (1,2,4,8,16) whilst food supply grows at an arithmetic rate (1,2,3,4,5). However, people have been susceptible to 'preventive' and 'positive' checks, such as war and disease which prevented the population exceeding the world's resources. Many current Neo-Malthusian demographers, such as **Erhlich**, are alarmed, because Malthusian checks seem to be diminishing as health care and sanitation improve. Malthusian ideas are particularly appealing to Western governments, as they highlight too many mouths to feed, rather than an uneven distribution of resources. It was based on these ideas that the Club of Rome was established in 1968 by Western industrialists amongst others. They commissioned a report from **Meadows et al** which was published as *'The Limits to Growth'* in 1972.

Boserup Thesis (1965): Instead of too many mouths to feed, Boserup emphasised the positive aspect of population growth. This, in simple terms, is that the more people there are, the more hands you have to work. She argued that population increase not the amount of food grown is the independent variable. So, as population increases, more pressure is placed on the existing agricultural system which stimulates invention and involution (intensification and technological improvements). This theory is unproven, but it challenges the popular view by asserting that **necessity is the mother of invention**.

Both theoretical arguments need to be well understood and you must be aware of their limitations and faults. As with everything, real world examples should be used to illustrate them.

Stabilisation Model ('S' curve): This follows the ideas of the DTM over time, a low base population is followed by a rapidly expanding one that finally stabilises and evens out at a new high population level.

Population Crash Model ('J' curve): This follows the ideas of Malthus in so far as the population rapidly expands, reaching the carrying capacity of the land. As the population exceeds this, famine and disease play havoc with the population and it crashes.

9 Population and resources

REVISION SUMMARY

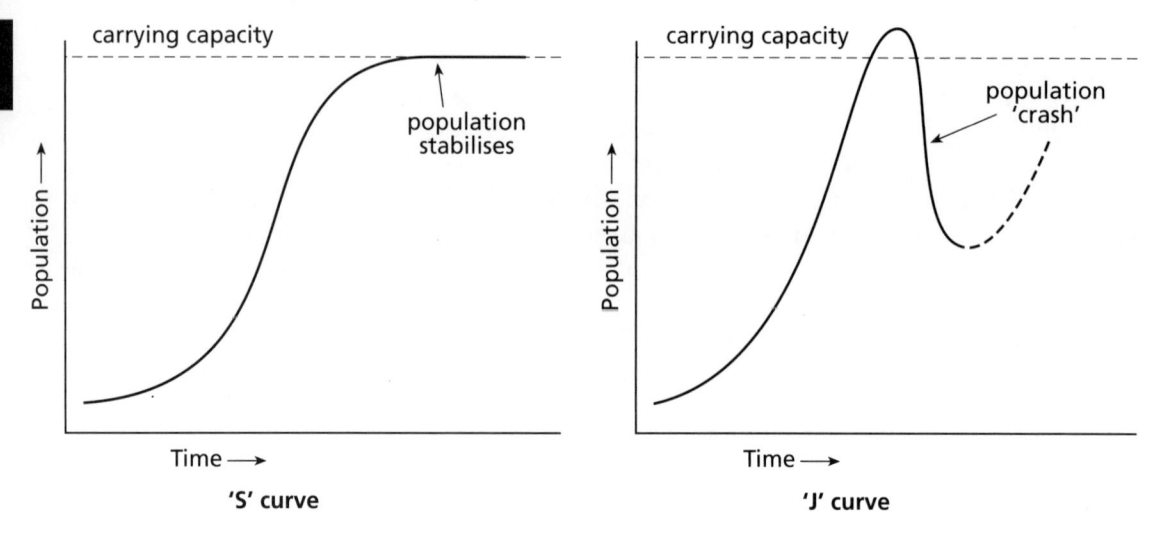

'S' curve

'J' curve

Resource/population relationships: These must be viewed with Malthus and Boserup in mind. Two crucial definitions here are **optimum population** (the number of people there are in relation to a given situation that produces the maximum return per person), thus under- and overpopulation can be calculated, and **carrying capacity** (the maximum number of users that can be sustained by a given set of resources). Candidates must not forget the crucial role that technology plays, because this can increase the carrying capacity by changing the resource base.

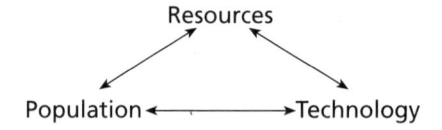

Resources ↔ Population ↔ Technology

Ageing societies
This is an issue of increasing importance as Developed World countries reach the final stage of the DTM. In 1985, 15.1% of the UK's population was over 65. By 2050 AD it is projected that 29% will be of pensionable age. This means that the **dependency ratio** will increase. How this affects actual places might be illustrated through retirement centres such as Bournemouth.

Expanding populations
The problem here is not an ageing population, but one where the massive increase in surviving children (not necessarily an increasing birth rate) is causing the rate of population growth to accelerate. This in turn results in often impoverished nations struggling to provide enough resources (food, housing, employment etc.) for the people to live, a situation often described as **'running to stand still'**. This is compounded by a high dependency ratio, this time caused by many children who are being supported by the smaller working population. Generally, most emphasis is placed on reducing the birth rate although in certain regions of countries, particularly cities, population growth can be accelerated by migration.

Migration
This, in many ways, is a sub-topic within population. You need to be able to define the term, understanding that this in itself is difficult because of the question of permanency. There is a distinction between mobility (travelling distances from a home base i.e. commuting) and migration, implying a more permanent move. It is necessary to study migration, as this is the most rapid catalyst of population change, due to either out-migration (emigration) or in-migration (immigration).

Some models you should know are:

EG Ravenstein (1885): This is the classic model in which several laws of migration outline some general characteristics of migrants as well as migration.

Population and resources 9

Distance decay models that link the relationships between the size of a settlement, its distance from potential immigrants and the total number of actual migrants. The most widely used model of migration is the **push/pull model** where people may be forced away from their place of origin (e.g. Rwanda's civil war or the Balkans' ethnic cleansing), or attracted to their destination (e.g. the lure of a possible job for a Turk to present day Germany). There are other models, such as Stouffer's idea of **intervening opportunities**, and although you may not need to know all the models, you must have a good theoretical background.

Having learnt migration in itself, you must realise the consequences of it and what spatial outcomes occur because of it (e.g. counterurbanisation in Developed Countries or urbanisation in Less Developed Countries). Remember too the self-reinforcing nature of migration: most immigrants are young, so not only does the population rise because of their immigration but also the migrants themselves are more fertile, so increasing the population still further.

Population policy

You will, of course, need in-depth case studies to back your arguments and ideas, for example China's one child policy, or the success story of the state of Kerala in India. You need to be able to say what is happening, where and why, and whether the policies are successful, considering the moral issues at the same time.

REVISION SUMMARY

If you need to revise this subject more thoroughly, see the relevant topics in the *Letts* A level *Geography Study Guide*.

9 Population and resources

QUESTIONS

1. Study Fig. 1 which shows percentage population change, by type of district, in England and Wales for the period 1961-1986.

    (a) (i) Describe the population changes which have occurred in Inner London between 1961 and 1986. (3)

    (ii) Suggest reasons for the high rates of population change in Inner London between 1966 and 1981. (4)

    (b) (i) Using data from Fig. 1, complete Fig. 2 (opposite) to show the population shift for the remoter rural districts for 1976-81 and 1981-86. (2)

Type of District	1961–66	1966–71	1971–76	1976–81	1981–86
Inner London	–8	–19	–20	–16	–2
Outer London	–1	–2	–7	–5	0
Principal cities	–8	–8	–11	–9	–4
Cities	1	–2	–3	–3	–3
Industrial	8	7	5	2	0
New towns	23	19	15	14	9
Remoter towns	8	10	12	8	9
England and Wales	6	4	1	1	2

Fig. 1

Use the formula: Population shift = (district change) – (change in England & Wales).

    (ii) Suggest reasons for the population shifts in: (8)

    1. principal cities;

    2. remoter rural districts.

    (c) For the whole period 1961-1986, explain the population changes affecting: (8)

    (i) industrial districts;

    (ii) new towns.   *ULEAC*

2. (a) Fig. 3 (opposite) shows a partly-completed diagram of the Demographic Transition Model. The model is based on the experience of Developed Countries.

    (i) name variable 'A', and variable 'B'; (2)

    (ii) specify the units that should be named at 'C'; (1)

    (iii) state the term usually used to describe the shaded area (Term 'D'); (1)

    (iv) state the descriptive names for Stage 2 and Stage 3. (2)

    (b) Fig. 4 (opposite) shows three graphs. One of the graphs is a more appropriate summary of demographic changes in **Developing Countries** since 1950 than the other two. Identify the graph that is more appropriate and justify your choice. (3)

QUESTIONS

Fig. 2

Population shift
+20
+15
+10
+5
0
-5
-10
-15
-20
61/66
66/71
71/76
76/81
81/86
● Remoter rural districts
○ Principal cities

Fig. 3

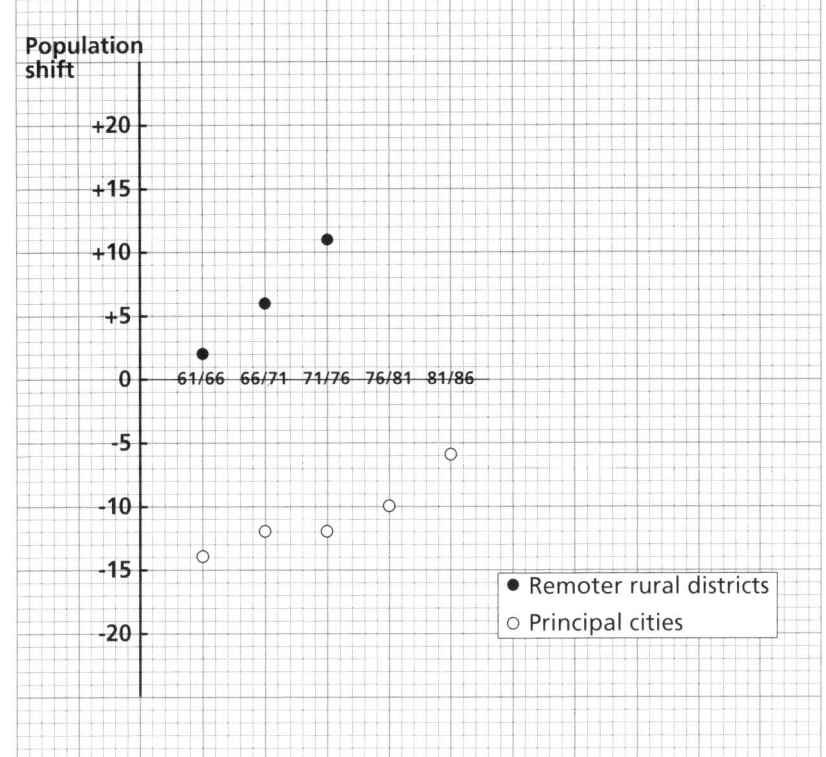

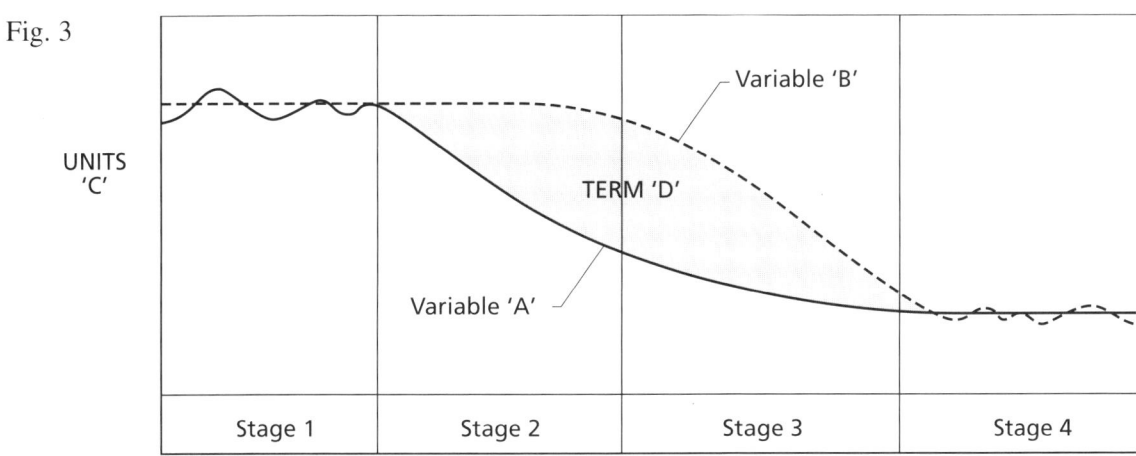

UNITS 'C'
Variable 'B'
TERM 'D'
Variable 'A'
Stage 1
Stage 2
Stage 3
Stage 4

Fig. 4

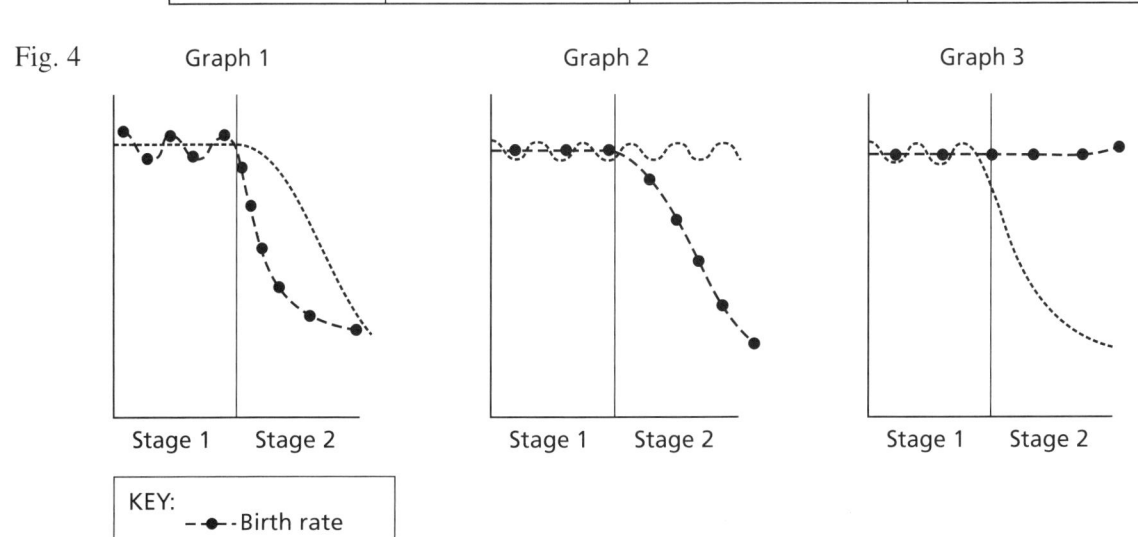

Graph 1
Graph 2
Graph 3
Stage 1
Stage 2
Stage 1
Stage 2
Stage 1
Stage 2
KEY:
Birth rate
Death rate

9 Population and resources

QUESTIONS

(c) With reference to specific examples explain why, in recent decades, the natural increase of population in Developing Countries has been higher than the increase of population in Developed Countries. (9)

(d) By reference to examples explain why, within regions of sparse population, there are areas where the population density is high. (7)

Oxford

3 Many governments have policies which favour either the growth or limitation of population. With reference to such countries describe the population policies that have been adopted and the reasons that underlie the adoption of these policies. (25)

Oxford and Cambridge

4 Study Fig. 5 which shows the Malthusian model of population growth.

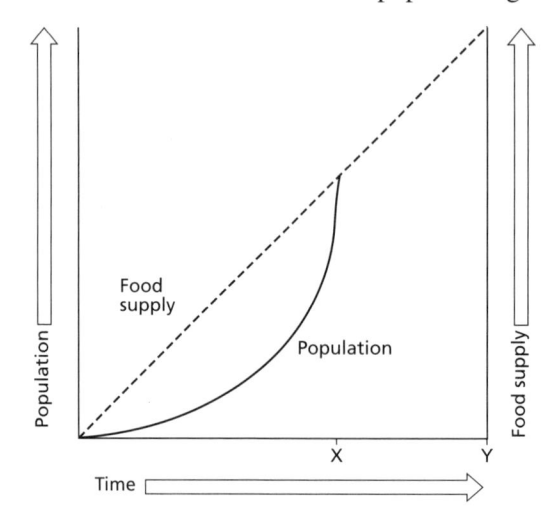

Fig. 5

(a) (i) Explain why:
1. population might increase at a geometric rate;
2. food supply might increase at an arithmetic rate. (4)

(ii) On Fig. 5 complete the probable curve of population for the time period X to Y. (2)

(iii) Justify the line you have drawn in (a) (ii). (3)

(iv) Suggest the limitations of the Malthusian model of population growth. (4)

(b) How might a country's food supply be increased?

(i) in the short term;

(ii) in the long term. (6)

(c) Why might a country

(i) encourage population growth?

(ii) discourage population growth? (6)

ULEAC

Industry 10

The definition of industry in its widest form encompasses all economic activities. These can be divided into four different classes, namely, **primary** (farming, fishing, mining and forestry), **secondary** (manufacturing industry), **tertiary** (services, retailing, transport and administration) and **quaternary** (hi-tech and information services). This section will look more at manufacturing industry: agriculture and tourism are dealt with separately.

You need to know what affects the location of industry. The many influences include: markets, transport, energy supplies, (raw) materials, chance, decision maker's preferences and perceptions etc. **Government policies**, industrial inertia and (dis)economies of scale including agglomeration and linkages, can also help to explain the distribution of industry.

REVISION SUMMARY

Theories of industrial location
These stem from the classical **Least Cost Location Model** (Weber 1909), and a thorough appreciation of the role of transport costs and the nature of a commodity (e.g. bulk) is necessary. This also includes the Material Index and the Varignon frame. **Maximum Revenue** (Losch), Rawstron's **Spatial Margins to Profitability** and the **Behavioural Matrix** theories have tried to refine the theoretical explanations for industrial locations.

Changing nature of industry in the Developed World
Includes the impact of **de-industrialisation** and **restructuring** of industry (including Government policy) and the move away from manufacturing to service based industries. Recent changes have affected people's lives and the environment through factors such as pollution. Knowledge of industrial dispersal to rural and suburban areas, especially that of **footloose** and hi-tech industry is required.

Industrialisation in the Developing World
You must have an understanding of **Rostow's Model of Economic Development**, so that industrialisation in Newly (Nearly) Industrialised Countries such as the Pacific Rim, and Least Developed Countries can be put into context.

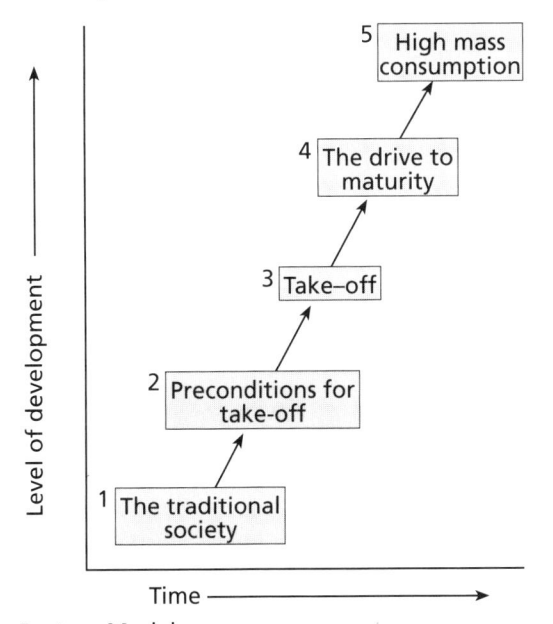

Rostow Model stages:
1. The traditional society
2. Preconditions for take-off
3. Take-off
4. The drive to maturity
5. High mass consumption

Approximate date of reaching a new stage of development

Stage Country	2	3	4	5
UK	1750	1820	1850	1940
USA	1800	1850	1920	1930
Japan	1880	1900	1930	1950
Venezuela	1920	1950	1970?	—
India	1950	1980?	—	—
Ethiopia	—	—	—	—

Organisation of industry and firms
Trans- or multi-national corporations are becoming more influential on the location of industry. New industrial practices of 'Just in Time' as opposed to 'Just in Case', and the life cycle of a product (R and D to mass production) also have some influence.

10 Industry

REVISION SUMMARY

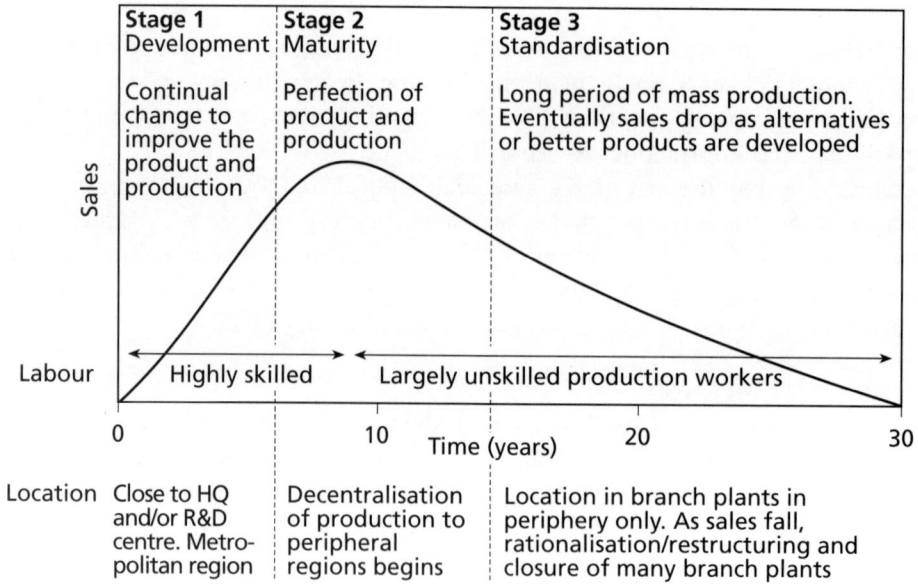

	Stage 1 Development	Stage 2 Maturity	Stage 3 Standardisation
	Continual change to improve the product and production	Perfection of product and production	Long period of mass production. Eventually sales drop as alternatives or better products are developed
Labour	Highly skilled	Largely unskilled production workers	
Location	Close to HQ and/or R&D centre. Metropolitan region	Decentralisation of production to peripheral regions begins	Location in branch plants in periphery only. As sales fall, rationalisation/restructuring and closure of many branch plants

Life cycle of a product

If you need to revise this subject more thoroughly, see the relevant topics in the Letts A level Geography Study Guide.

Detailed case studies are necessary to illustrate these concepts. A variety of examples are needed from both market (e.g. hi-tech) and raw material (e.g. steel) orientated industries and from the Developed and Developing Worlds. Some syllabuses require knowledge of industry within a Socialist command economy.

Industry 10

QUESTIONS

1 'Just as the forces of the past are felt in present day industrial patterns, so the greatest influence on the future location of industry is likely to be its present location.' Discuss using detailed examples. (25)

Oxford and Cambridge

2 (a) Outline Weber's theory of industrial location. (10)

   (b) To what extent is Weber's theory less applicable today? (15)

AEB

3 Study Fig. 1, which shows changes in percentage employment by employment sectors.

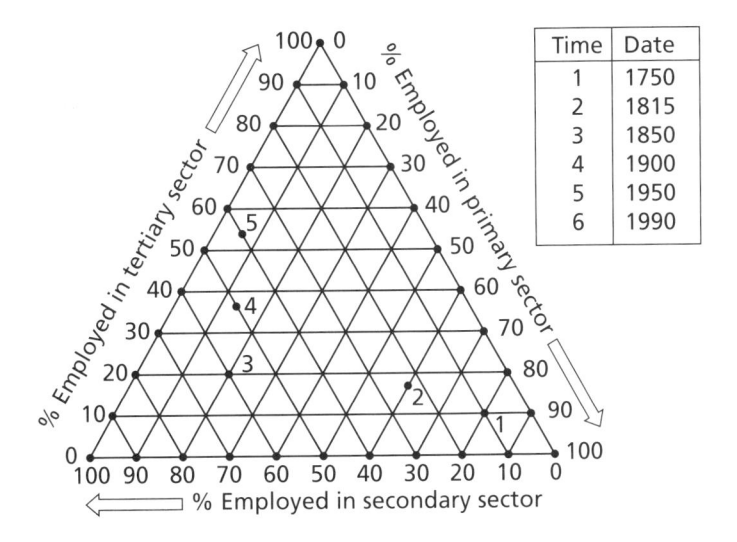

Fig. 1

Time	Date
1	1750
2	1815
3	1850
4	1900
5	1950
6	1990

   (a) (i) Using Fig. 1 state the percentage values for the three employment sectors in:

   1750   Primary   Secondary   Tertiary

   1900   Primary   Secondary   Tertiary (2)

   (ii) On Fig. 1, plot for 1990 the percentage values for the three employment sectors: Primary = 4%, Secondary = 36%, Tertiary = 60% (1)

   (b) Suggest reasons for the changes in employment:

   (i) in the primary sector between 1750 and 1990; (4)

   (ii) in the secondary sector since 1990. (4)

   (c) (i) Suggest what changes may have taken place in the types of employment in the tertiary sector. (3)

   (ii) Explain these changes. (4)

10 Industry

QUESTIONS

(d) Since 1990, in country X, another employment sector has developed. It is termed the 'Quaternary sector'.

   (i) Explain the meaning of the term 'Quaternary sector'. (3)

   (ii) Account for the recent growth of this sector. (4)

                                                                                                           ULEAC

4 (a) An entrepreneur has decided to set up a small engineering factory.

Fig. 2 shows information about four possible locations for the factory A, B, C, and D. The graph shows the cost curve and the revenue curve from which can be derived the spatial margins of profitability. Costs and revenues are given per unit of production.

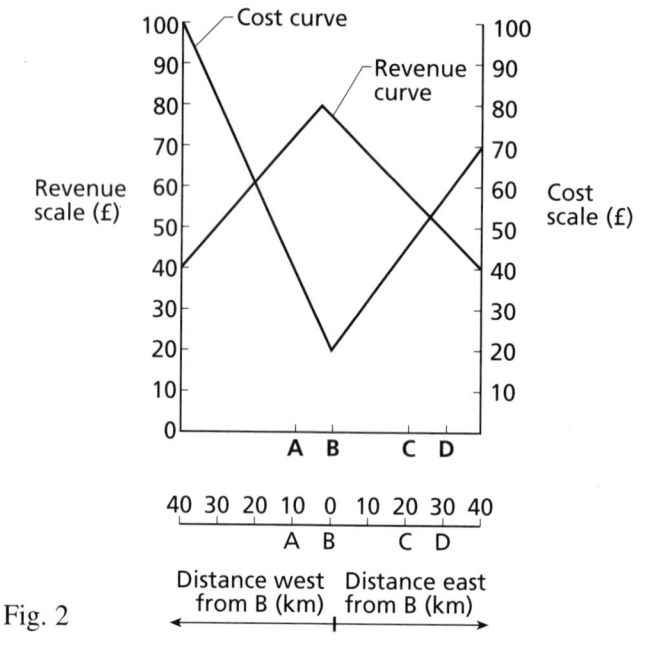

Fig. 2

Study Fig. 2 and answer the questions that follow.

   (i) State the distance of the spatial margin of profitability to the east, and that to the west of location B. (2)

   (ii) In terms of profit, state which of the terms A, B, C, and D can be described as optimal and which can be described as sub-optimal. (3)

   (iii) With reference to Fig. 2, calculate the profit or loss per unit of production for locations A, B, C and D. (4)

(b) Briefly explain, with reference to **a specific example**, the difference between a factory and a firm. (3)

(c) With reference to specific examples, explain why multi-national firms in the motor industry have located different parts of their operations in different parts of the world. (13)

                                                                                                          Oxford

Industry 10

5 (a) Study Fig. 3.

Region	Employment	Location Quotient
Yorkshire and Humberside	26,086	7.21
Scotland	7,617	1.97
North West	2,857	0.61
East Midlands	1,391	0.47
South West	861	0.27
West Midlands	854	0.21
Northern	839	0.39
Rest of South East	654	0.09
Wales	562	0.37
London	397	0.06
East Anglia	278	0.19
Total	42,396	1.00

Fig. 3 *The spatial distribution of the woollen textile industry in Great Britain, 1984*

    (i) Explain what the Location Quotient tells us about the woollen textile industry in Yorkshire and Humberside, Wales and the Rest of South East England. (7)

    (ii) Since 1945 some new woollen textile plants in the North, North West and Yorkshire and Humberside regions were established as a result of government regional policy initiatives. Do you think that the government should offer such initiatives to support manufacturing industry? (8)

(b) With reference to both transnational corporations and the internationalisation of capital, describe the global restructuring of one major manufacturing industry. (20)

NICCEA

11 Agriculture and food

REVISION SUMMARY

Food is crucial for our health and to support our population numbers, and so how we organise its production is of fundamental importance.

You need to know about types of agricultural systems, their distribution and how they can be classified by their characteristics such as **peasant/commercial**, **shifting/sedentary**, **intensive/extensive** etc. You must consider how the physical environment influences agriculture through a wide range of variables including **climate**, **relief**, **soil** and **hydrology**. You also need examples to show these. McCarty and Lindberg's **optimum and limits** model states there is an ideal location for every type of farming due to physical and environmental differences.

You need to study **ecosystems**, **bio-diversity**, **stability** and the **nutrient cycle**, (see the earlier section on ecosystems) and the influence and impact of human activity.

The human influence on farming is also critical. See von Thünen's **isolated state theory** (1826), and its modifications, in explaining economic influences such as the location of a market: you need to understand its assumptions and limitations.

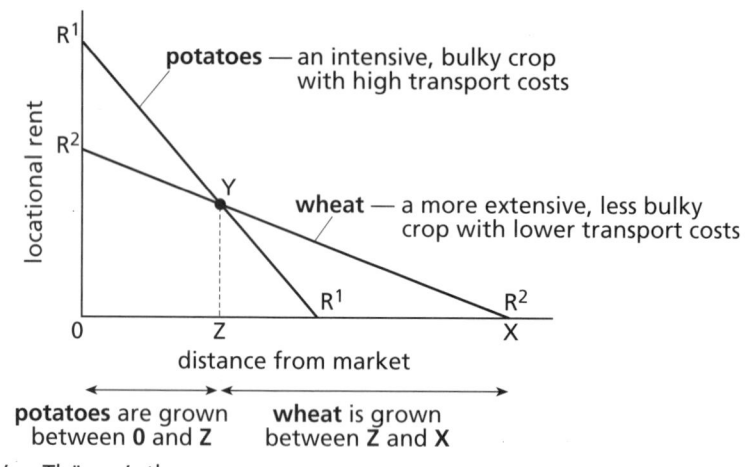

Simplified graph after Von Thünen's theory

The importance of people as decision makers is examined through the Behavioural Matrix, and our ability to develop new techniques such as biotechnology and high yielding varieties (HYV) has also contributed to increasing agricultural output (e.g. the **Green Revolution**).

Governments modify agriculture as do supra-national schemes such as the European Union's Common Agricultural Policy. State ownership of the farming system such as the Sovkhoz in Russia leads to different patterns than freehold arrangements. Land tenure differences need to be considered, for example consolidation in France or fragmentation in Ethiopia. Population densities and numbers necessitate agricultural evolution and change such as shifting cultivation to bush fallowing in many parts of West Africa.

You must also consider the impact of agriculture on people and the environment. Farming practices can change natural systems and lead to **land degradation** through overgrazing, salinisation, deforestation and soil erosion. This can interfere with hydrology, and changes in microclimate occur. You must also bear in mind the **cultural heritage** of the countryside and our perception of it, plus the **destruction** or **modification** of habitats (e.g. loss of water meadows): agriculture can become an issue of conflict.

Agriculture should be put in the context of the **nutritional requirements of people** (calories per day). The distribution of food-related diseases, such as rickets and beriberi and **malnutrition** shows the (uneven) distribution and production of food. This leads to **famine** and its causes, both physical and, arguably more importantly, human factors. Attempts to alleviate hunger, including **food aid**, should be known.

You need detailed case study material to illustrate these concepts.

If you need to revise this subject more thoroughly, see the relevant topics in the Letts A level Geography Study Guide.

Agriculture and food 11

1 (a) Define the term *shifting agriculture*. (3)

(b) Study the diagrams (Fig. 1) below which illustrate the relationship between soil productivity and length of fallow period experienced by shifting agriculture in savanna regions.

Fig. 1

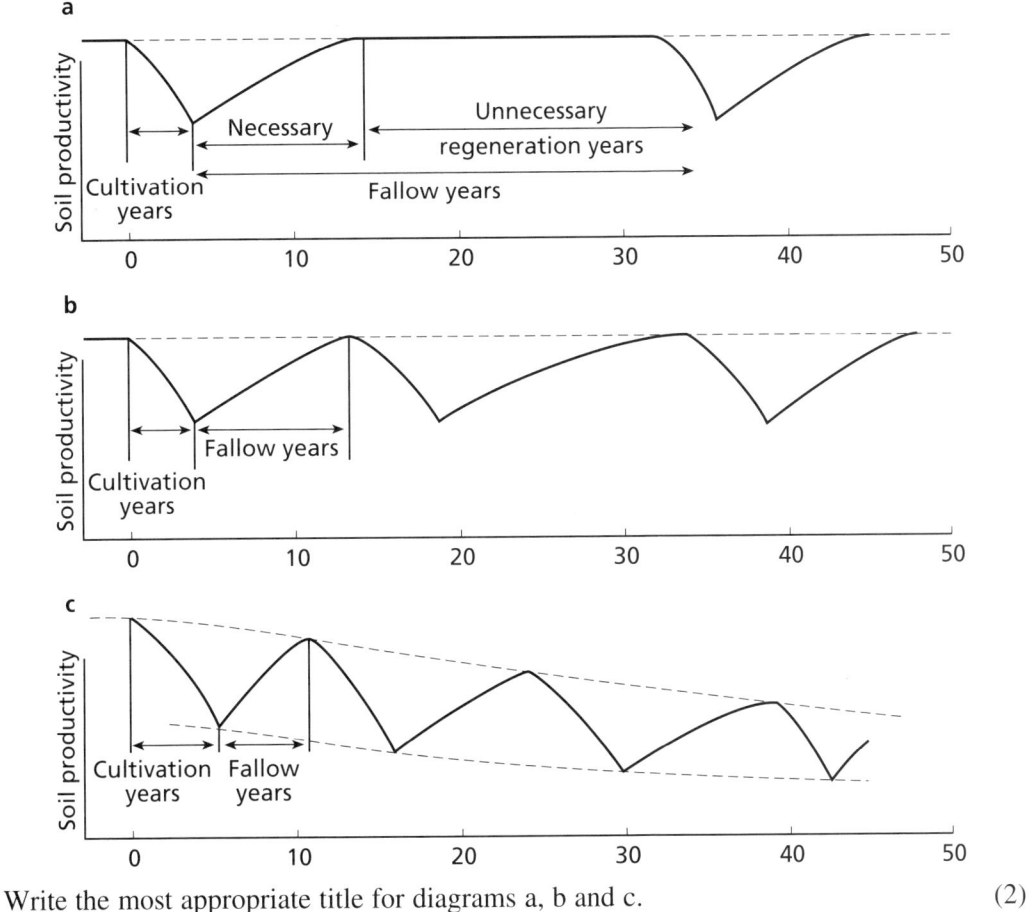

a

b

c

Write the most appropriate title for diagrams a, b and c. (2)

Optimal   Underutilisation   Overutilisation

(c) List two factors which might lead to a fall in soil productivity in a savanna region. (2)

(d) Suggest three ways by which the food supply in a savanna region may be increased whilst maintaining soil fertility. (6)

(e) How might the frequency of rotational fallow vary around settlements in savanna regions? (2)

AEB

2 (a) Describe the main features of 'plantation agriculture'. (10)

(b) Evaluate the benefits and the problems associated with this farming system. (15)
ULEAC

11 Agriculture and food

QUESTIONS

3 'Farmers may seek satisfactory rather than optimal decisions and these will be related to attitudes, perceptions and values, as well as past experiences.' Using examples, explain how these factors can influence agricultural land use patterns. (25)

Oxford and Cambridge

4 (a) Rice is the staple grain crop of South East Asia, and its production is closely associated with the region's monsoon climate.

Describe and account for the typical pattern of temperature and rainfall in a rice growing area of South East Asia. Outline the advantages of this pattern to a peasant farmer. (12)

(b) Referring to specific examples:

(i) describe the main features of the so-called 'Green Revolution' which have changed peasant farming in many parts of the Developing World; (8)

(ii) discuss the extent to which the Green Revolution has been successful in improving agricultural output. (5)

SEB

5 (a) Use the advertisement below (Fig. 2) to consider the main arguments that the American cattlemen put forward to prove to the public that they care for the environment and that 'every day is Earth Day'. (7)

Fig. 2

EVERY DAY IS EARTH DAY FOR AMERICAN CATTLEMEN

The American cattleman is still hard at work out there. We're out West, down South, up North, back East and in the Midwest heartland. Sometimes we wear a 10-gallon hat with a suit and tie...ride a pickup truck instead of a horse. We've even traded our dusty ledger books for computers. But some things haven't changed — we still do business on a handshake. And we still work hard to care for the land that is our livelihood and our future.

Cattlemen own or manage more land in this country than any other industry. We raise our cattle primarily on the hundreds of millions of acres of U.S. land unsuitable for crop production. But that land produces renewable resources like grass and forage. We use those resources, as well as feedgrains and harvest roughage not edible by humans, to produce a healthful, nutritious food for humankind. We are stewards of the miraculous cycle of *sun* to *grass* to *cattle* to *human food*.

We take our responsibility for natural resources seriously. We invest millions of our own dollars every year to maintain and improve the land and water we use. It's an investment in food for the 21st century...and an investment in future generations of cattlemen.

In the process of raising cattle, we also help Mother Nature. Cattle grazing helps strenghten and replenish vegetation — it's a lot like mowing your lawn.

Wildlife also benefits from our care of natural resources. Cattle grazing keeps plant life fresh and succulent, providing a healthy habitat for many species of wildlife which share our lands. And when nature isn't kind — when there are blizzards, windstorms, fire or drought — we help minimize nature's damage.

Cattlemen are partners with the land — we live and work close to the Earth every day. We're proud to celebrate the 20th Anniversary of Earth day, April 22 — because every day is Earth Day for American Cattlemen.

National Cattlemen's Association

The Beef Promotion and Research Board

Agriculture and food 11

QUESTIONS

(b) Choose an example of a named farming system you have studied in any part of the world.

   (i) Draw an annotated diagram to summarise its main features.

   (ii) Describe and explain how the natural environment both promotes and constrains agricultural practice in your chosen system. (18)

ULEAC

6 'Natural climactic variations can no longer be blamed for modern famines. Famines are man-made.' Discuss with reference to specific places you have studied. (25)

Oxford and Cambridge

12 Tourism and leisure

REVISION SUMMARY

Globally, tourism is now the world's second largest activity in terms of money generated. Since the 1960s, there has been an explosion in leisure both in the home (centred on electronic equipment) and outside the home, as access to cars has increased people's mobility.

You need to know about the evolution and historical development of tourism. Tourism has flourished since the late eighteenth century, linked in particular to **rising incomes**, **increasing leisure time** and **improvements in transport technology** that continue to increase mobility. For example, the development of seaside resorts on the south coast of England was closely linked to the development of the railways. Further shrinkage of distance through the growth in air travel has allowed tourism to expand globally.

You must be aware of human influences on tourism, and its impact on people. Governments have often been instrumental in promoting tourism (e.g. Jamaica) as it can aid economic development and positively influence investment in the infrastructure. The downside of this is that **overdependence** can occur. The stability of the political climate can also affect the flow of tourism, as seen in the Gambia in 1995. Careful management and controls are often necessary to help tourist centres flourish, to help both the local environment and inhabitants to cope with the impact of tourism, e.g. the creation of the American National Parks, including Yosemite, which operate a 'quota' system to control entry.

Tourism obviously brings advantages to people (i.e. jobs) and countries (foreign currency), but it also brings problems such as a dilution of cultural heritage (e.g. the 'Lager Louts and Chips' image of the Costa del Sol). The presence of the tourists, both as individuals and large groups affects popular areas substantially. Tourism is often seen as a 'fashion' industry, hence places are vulnerable to consumer fads.

Increasing impact ↓

Type	Numbers	Impact	Who, me?
Explorer	Very few	Accepts local conditions	Explorers
Offbeat	Small numbers	Revels in local conditions	'Across the Sahara by bus'
Elite	Limited numbers	Demands Western amenities	The international Hilton set
Early mass tourism	Steady flow of numbers	Looks for Western amenities	Professional middle classes
Mass package	Massive arrivals	Expects Western amenities	Everybody and their mothers!

A model of tourism progression

NB: This model can be applied to an individual location through time, or to classify different places at a given moment in time.

Leisure analysis is a useful tool to explain this. The **density** of people at a resort is less important than **crowding** (a qualitative judgement) and it also depends on the type of leisure, and person, involved, such as the differing wishes of those on an 18 to 30 Club holiday compared with someone fell walking. The recreational carrying capacity of an area will vary greatly and can be judged by physical numbers of people, ecological damage and perceptions.

Tourism and leisure 12

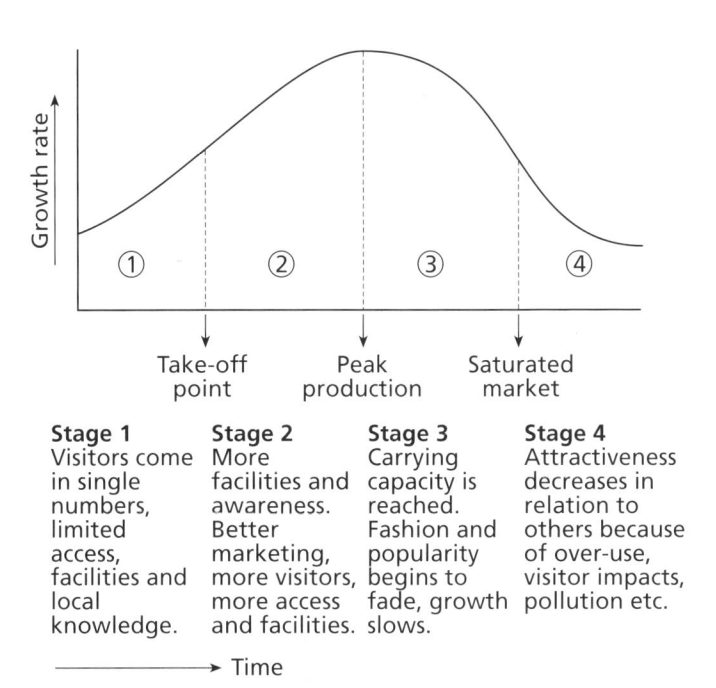

REVISION SUMMARY

Stage 1 Visitors come in single numbers, limited access, facilities and local knowledge.

Stage 2 More facilities and awareness. Better marketing, more visitors, more access and facilities.

Stage 3 Carrying capacity is reached. Fashion and popularity begins to fade, growth slows.

Stage 4 Attractiveness decreases in relation to others because of over-use, visitor impacts, pollution etc.

NB Planned 'instant' resorts such as Languedoc/Roussillon (France), Cancun (Mexico), Les Arcs (The Alps) do not fit this model.

Physical influences on tourism must also be considered. The physical quality (including climate) and amenity value of the land is often the stimulus for tourist development, for example exotic tropical beaches. However, as tourism develops it can have a detrimental negative feedback as the ecological carrying capacity is reached (e.g. footpath erosion, destruction of coral reef or litter). Thus, the very thing that people wish to see is destroyed by them. It is clear that management is essential to relieve the stress caused by tourism: the conflict over land use and its care includes both human and environmental factors.

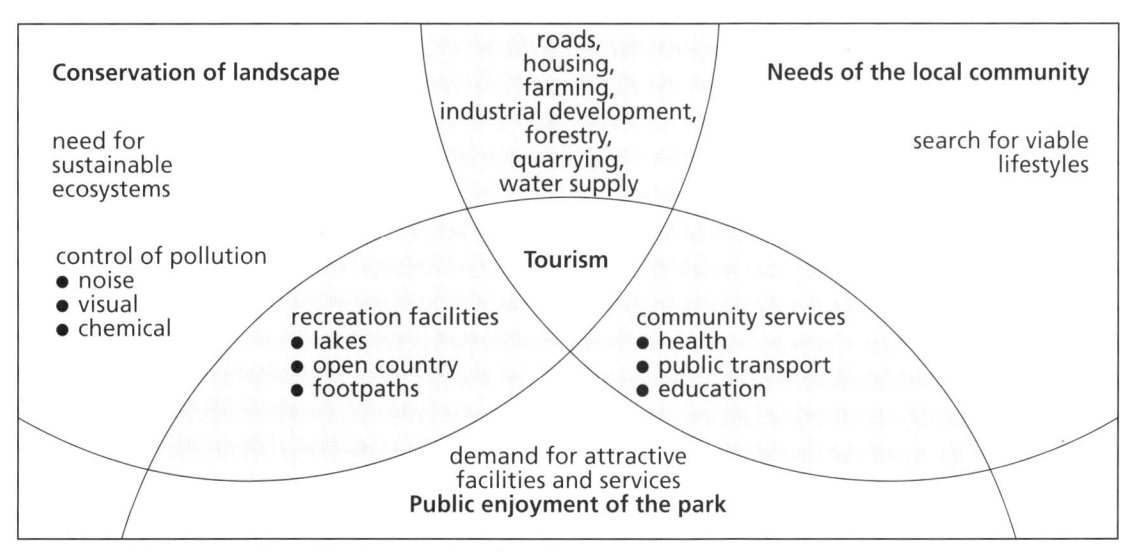

Some possible causes of conflict in a National Park

It is very important that these concepts are addressed at both a national and international scale and in both a Developed and Developing World environment.

If you need to revise this subject more thoroughly, see the relevant topics in the *Letts* A level *Geography Study Guide*.

12 Tourism and leisure

QUESTIONS

1 (a) Referring to Fig. 1 below, briefly summarise, in a form suitable as a handout for the local press, the changes that may result from the La Filette coastal development proposals. (7)

(b) Evaluate the potential environmental and economic impacts of the proposals. You should consider the impacts on landform processes, ecosystems, and existing and future economies. (18)

ULEAC

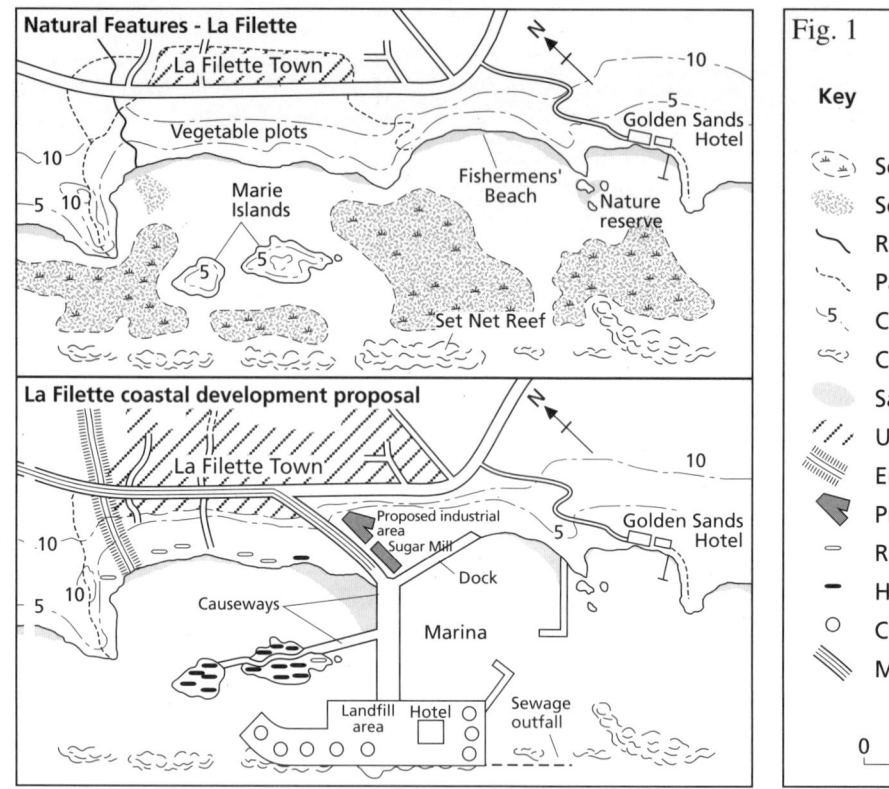

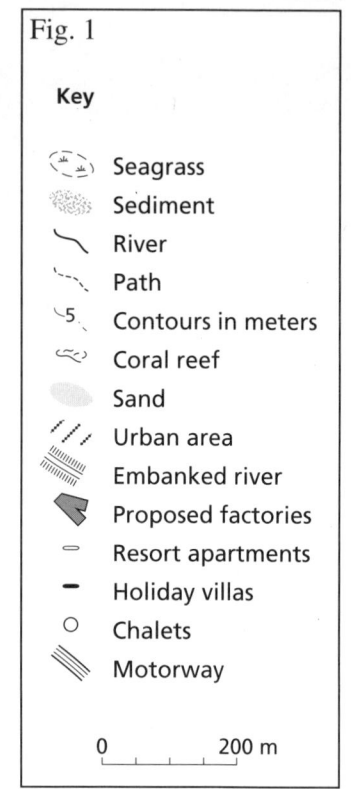

Fig. 1

Natural Features - La Filette

La Filette coastal development proposal

Key:
- Seagrass
- Sediment
- River
- Path
- Contours in meters
- Coral reef
- Sand
- Urban area
- Embanked river
- Proposed factories
- Resort apartments
- Holiday villas
- Chalets
- Motorway

0   200 m

2 (a) For any named area or areas you have studied, outline the impact of tourism on the local population, communities and cultures. (9)

(b) Consider why tourism is frequently seen as an unstable base for economic growth and prosperity at the local scale, yet continues to be a major world industry. (16)

UCLES

3 (a) Show, with examples, how and why Britain's seaside resorts developed in the nineteenth century and early twentieth century. (15)

(b) Outline some of the problems created by this rapid development. (10)

Oxford and Cambridge

4 With reference to specific examples from the case study areas, discuss:

(a) the relative importance of the physical environment and local culture as attractions for international tourists; (12)

(b) variations in and among the case study areas in the origin of international tourists; (6)

(c) the factors which may discourage the growth of international tourism. (7)

NEAB

Spatial inequality, regional disparities and development 13

REVISION SUMMARY

As places have different characteristics, so different levels of growth occur. Therefore, inequalities or disparities are naturally found between places. Areas that have similar characteristics are known as **regions**.

You need to know **the characteristics and measurements of spatial inequalities** which can be defined by welfare (number of doctors per capita), population (fertility rates) or economic factors (G.N.P. per capita). A number of different bodies such as the United Nations and the European Union have established criteria to measure 'wealth'. You must study the different models of economic development over time that help our understanding of spatial inequalities. A good appreciation of **Rostow** (see industrial section), **Myrdal** (multiplier effect and cumulative causation) and **Friedmann's** core-periphery model is necessary.

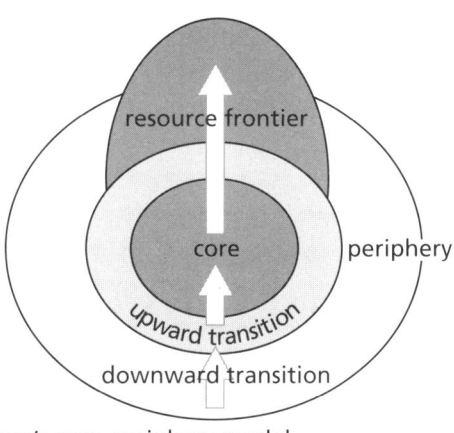

The main categories of Friedmann's core-periphery model

The belief that the Developed World exploits the Developing World and accumulates wealth is covered by the Marxist theories of **Frank**, and **Wallerstein's** theory alleging that the Developed World manipulates the world economy's market exchange in its favour is useful in illustrating global inequalities.

You need to study national government and supra-national bodies' (such as the Brandt Commission and the E.U.) attempts at developing and implementing **regional development strategies**. You should be able to assess their success or failure, and have a knowledge of the underlying reasons for their policies. Additionally, you need to be able to show how the relationship of trade and aid between the Developing World and the Developed World can demonstrate how inequalities can be reduced. **Schumacher's** approach of 'Small is Beautiful' outlining appropriate and sustainable development gives an alternative development pathway.

Finally, some syllabuses require a knowledge of ethnicity, religions and racial divisions in cities (see the settlement section) and the historical basis of these (e.g. apartheid).

These concepts need to be illustrated by detailed case studies at all scales from micro (within cities) to meso (inter-regional) and macro (global).

If you need to revise this subject more thoroughly, see the relevant topics in the *Letts A level Geography Study Guide.*

13 Spatial inequality, regional disparities and development

QUESTIONS

1 (a) Study Fig. 1 and and Fig. 2 below, and indicate what they say about the attitudes of some people from the Developed World to Developing Countries. (5)

(b) In terms of development processes, define:

  (i) the swash/backwash effect;

  (ii) the multiplier effect. (6)

(c) What is neo-colonialism? (4)

(d) With reference to specific areas, explain the measures that can be taken to reduce contrasts in levels of economic development at **either** the national **or** the international scale. (20)

NICCEA

Fig. 1

Rich must stop being foul to Africa's poor

A chilling proposal for increased pollution has come from the pen of the World Bank's chief economist, Lawrence Summers.

'I've always thought under-populated countries in Africa are vastly under-polluted,' Summers said in an internal memorandum leaked to the press. 'Shouldn't the World Bank be encouraging more migration of the dirty industries [to such countries]? I think the economic logic behind dumping toxic waste in the lowest-wage country is impeccable.'

The memo's publication led to the World Bank apologising on Mr Summers' behalf. But the man himself said that writing the memo he had 'tried to sharpen the debate on important issues by taking as narrow-minded an economic perspective as possible.'

In other words it had been an intellectual exercise. But if Summers had been sounding off only theoretically would any World Bank economist have broken the strong bonds of confidentiality that bind staff to leak the memo?

Cameron Duodo, The Observer, 02 March 1992

Fig. 2

Cartoonist: Balaram Thapa
Source: Worldwide Fund for Nature

2 Examine the arguments for and against government intervention aimed at reducing spatial inequalities. (20)

WJEC

Spatial inequality, regional disparities and development 13

QUESTIONS

3   For any **one** country in the Developed World:

   (a) Locate those regions which are receiving regional aid and explain why such aid has been focused on them. (15)

   (b) Evaluate the effectiveness of government assistance to one region of that country. (10)
   AEB

4   (a) Why is unbalanced economic development between regions within a country often seen as undesirable? (9)

   (b) How far do you agree with the view that such unbalanced development is solely the result of the working of market forces through time? (16)
   UCLES

5   (a) Outline one model of regional economic development (8)

   and

   (b) comment on the extent to which all models of regional economic development are only partly successful in explaining regional inequalities. (17)
   Oxford and Cambridge

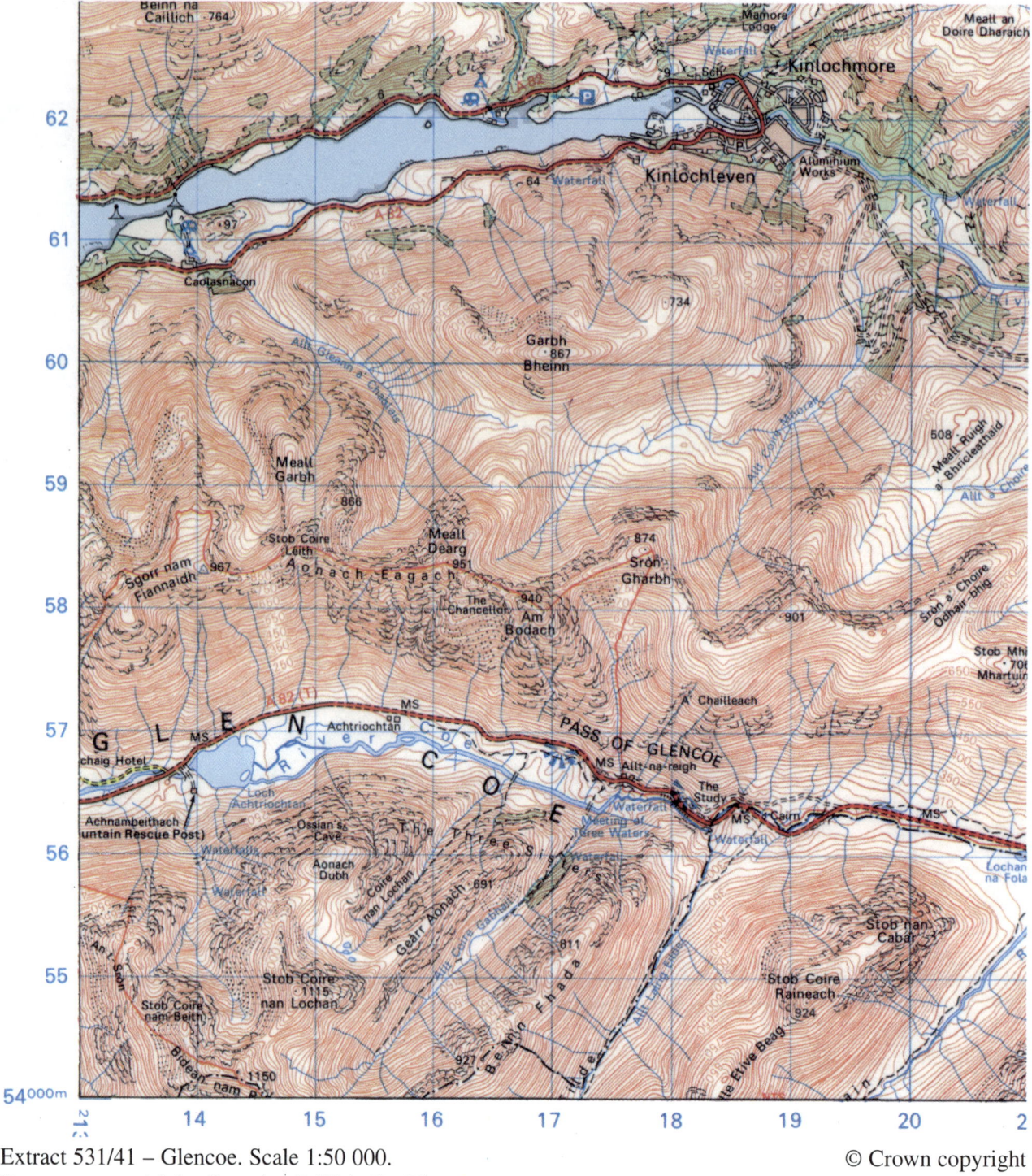

Extract 531/41 – Glencoe. Scale 1:50 000. © Crown copyright
Resource material for Question 2 (Unit 6) p35.

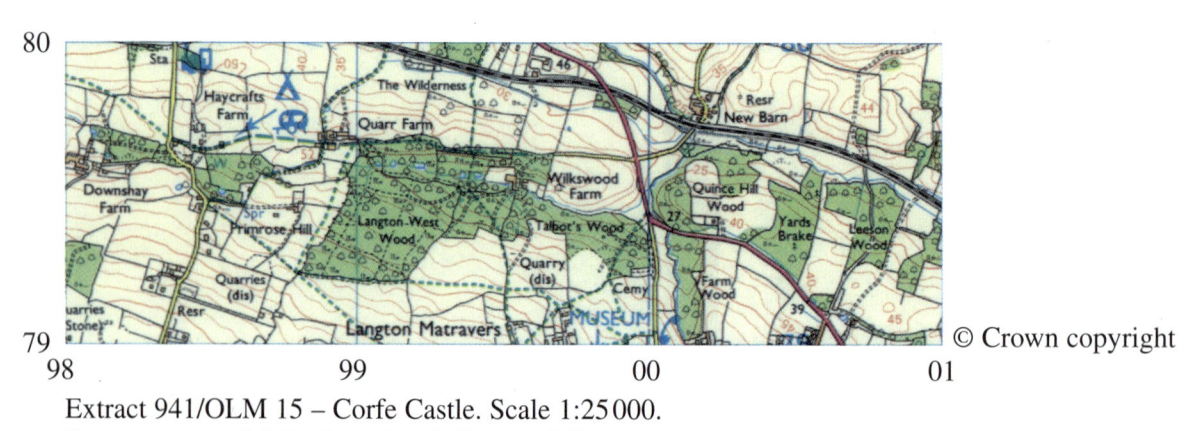

Extract 941/OLM 15 – Corfe Castle. Scale 1:25 000. © Crown copyright
Resource material for Question 1 (Practical Geography) p72.

Resource material for Question 2 (Unit 5) p32.

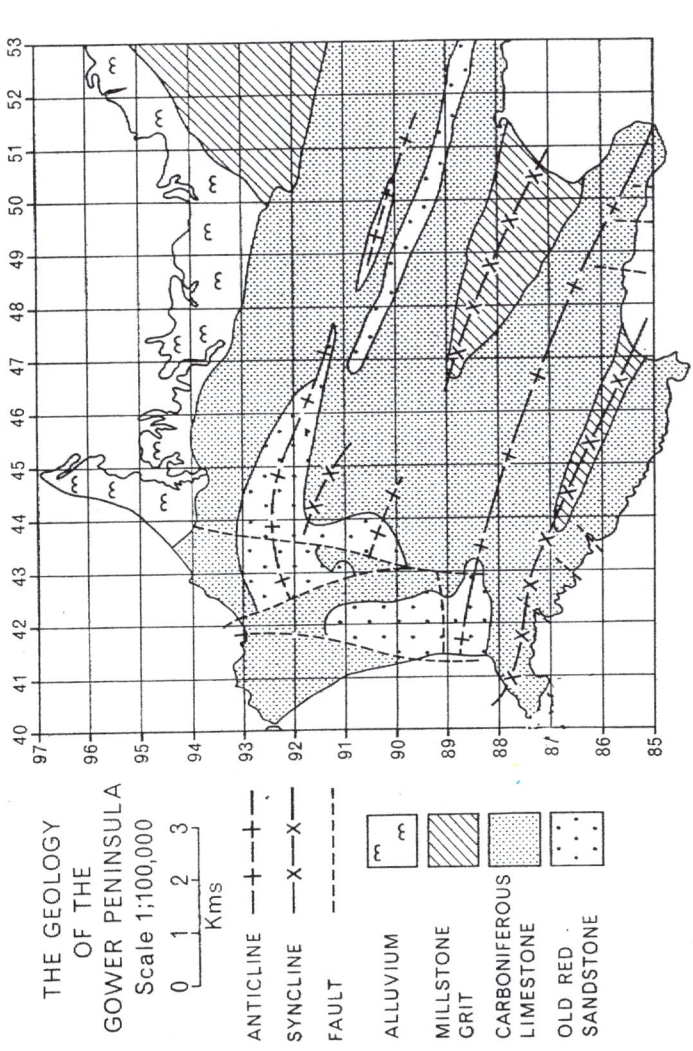

THE GEOLOGY
OF THE
GOWER PENINSULA
Scale 1:100,000

ANTICLINE —+—+—
SYNCLINE —x—x—
FAULT — — —

ALLUVIUM
MILLSTONE GRIT
CARBONIFEROUS LIMESTONE
OLD RED SANDSTONE

Extract 792/159 – Gower Peninsula. Scale 1:50000.

© Crown copyright

Practical Geography

QUESTIONS

1   Refer to OS 1 : 25 000 Map Extract No 941/OLM 15 (Corfe Castle), page 70.

You wish to undertake a microclimate study of Langton West Wood (992795) over the course of one year. The study will be based upon a transect from 992795 to 996792. Explain how you might plan your study, paying particular attention to:

   (i) the spatial sampling framework;

   (ii) the instruments you would require;

   (iii) the method you would use. (25)
   Oxford and Cambridge

2 **Either** Explain in detail how *you*, with two assistants, would establish the sphere of influence of a town of your choice that satisfies the following criteria:

   • a coastal holiday centre;

   • a population of less than 20 000.

   Or Explain in detail how *you*, with two assistants, would establish the degree of attraction of a town with the criteria shown above as a holiday centre. (25)
   modified Oxford and Cambridge

Answers

There are fully worked answers to at least one question per unit, plus examiner's tips on how to tackle the remaining questions.

1 THE ATMOSPHERE

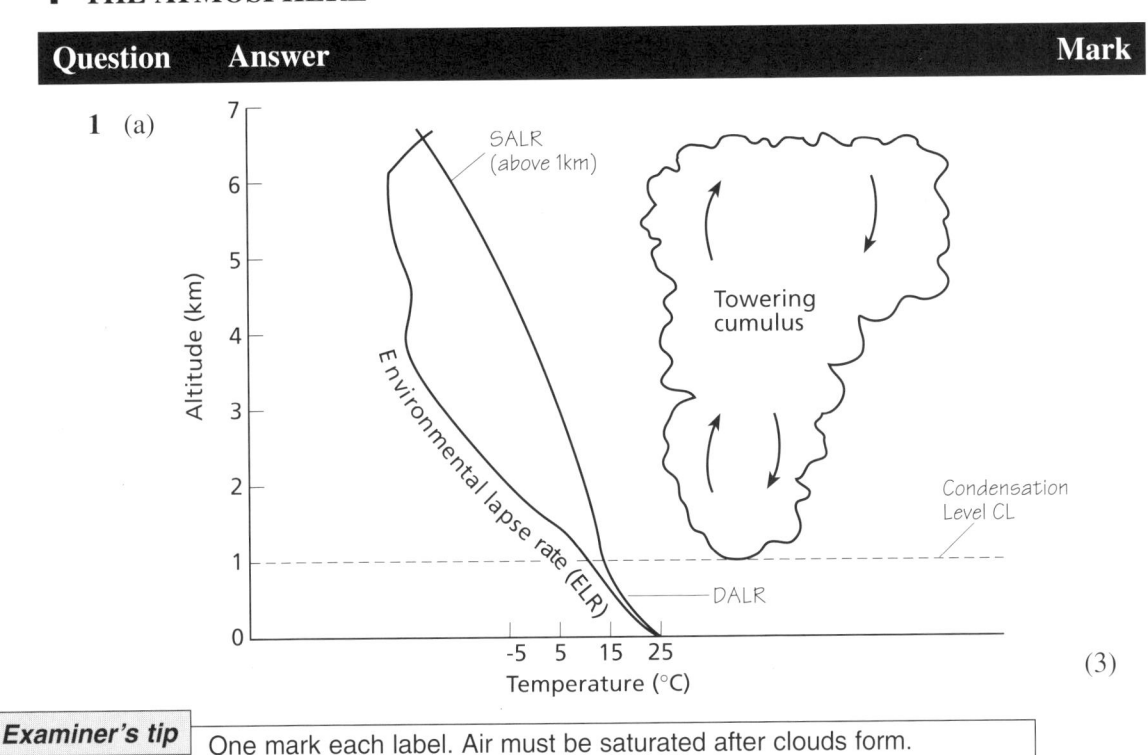

Question	Answer	Mark

1 (a)

[Graph showing Altitude (km) 0–7 on y-axis versus Temperature (°C) -5, 5, 15, 25 on x-axis. Labelled curves: Environmental lapse rate (ELR), SALR (above 1km), DALR. A towering cumulus cloud shape is drawn with arrows indicating circulation. A dashed horizontal line at 1km marks the Condensation Level CL.] (3)

Examiner's tip — One mark each label. Air must be saturated after clouds form.

(b) The ELR is the fall in temperature with increased height in the atmosphere. (2)

Examiner's tip — It pays to learn definitions – easy marks!

(c) The SALR has a slower rate of cooling than the DALR as latent heat is released by the air as water condenses from vapour. (2)

(d) (i) Prior to the cloud formation it has been sunny, resulting in heating and warm temperatures. Once clouds form there will be cumulus/cumulo nimbus type cloud cover with heavy rain; hail and thunder may result.

(ii) Stable air conditions will probably mean more hours of sunshine, better for working up a tan! Rain prevents many activities such as golf and sightseeing.

(2 + 2)

Examiner's tip — Only give two reasons for (ii). Think, why should stable, i.e. no rain, be an advantage?

2

Examiner's tip — All these answers require good, annotated diagrams; they will save writing, and clarify the processes. To score well your answer must be organised, and the theoretical concepts integrated into your account of the topic. Three topics have been selected, but hail and jet streams would need similar treatment. Focus on *processes* and try to include examples.

Answers to Unit 1

Question	Answer	Mark

Advection fog

A fog is produced where moisture condenses and reduces visibility to under 1 km. Advection is the movement of a fluid, such as air, in a horizontal, not vertical plane. Advection fog results when a warm, moist stable mass of air moves horizontally across a cooler surface. This results in cooling of the air, and as cold air can hold less vapour than warm air, the dewpoint temperature is reached and condensation results.

Formation of advection fog

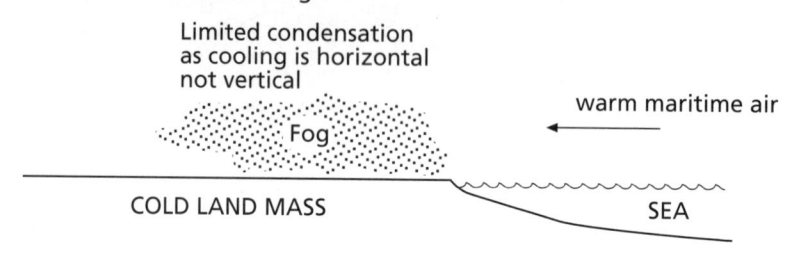

- Limited condensation as cooling is horizontal not vertical
- Fog
- warm maritime air
- COLD LAND MASS
- SEA

(8)

> **Examiner's tip** Give local or your own examples wherever possible. For example, you may have seen steam fog forming over a river during winter when cold air moves over water which is relatively warm, or seen a fog form in a coastal area as warm air is blown over the cooler sea.

Anabatic (up-valley) wind

Anabatic winds are local up-slope winds which occur in clear, sunny weather conditions within mountainous regions, for example, at Zermatt in Switzerland. These winds result from intense heating during the morning of partially vegetated, rocky mountain sides, which have a low albedo. Albedo is a measure of a surface's reflectivity. Dark soil will only reflect about 3% of incoming light, so bare rocky surfaces heat up rapidly. The warm ground heats the overlying air which then rises as it is less dense. This movement of air draws air up the slope from the valley because of the pressure gradient, producing a local wind.

Formation of Anabatic or valley wind

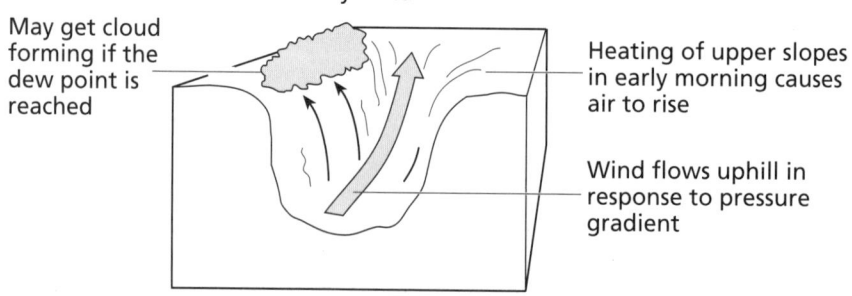

- May get cloud forming if the dew point is reached
- Heating of upper slopes in early morning causes air to rise
- Wind flows uphill in response to pressure gradient

> **Examiner's tip** Define terminology (e.g. albedo) when necessary.

This process is often associated with the development of clouds on the crest of the valley when the dewpoint temperature is reached. Under optimum conditions winds may reach 10-15 m/s^{-1}. Up-valley winds tend not to last very long as

Answers to Unit 1

Question	Answer	Mark

heating in the valley as the day proceeds reduces the pressure gradient. The updrafts produced by these anabatic winds have led to the development of hang gliding as a popular sport during the summer in places such as Chamonix in France. In the Alps, during the summer, the clear, cloudless mornings can be followed by dramatic thunderstorms and sudden rainstorms. The process may be reversed at night leading to cold winds blowing down from the mountains at night. These are known as katabatic winds and are more persistent. (9)

Orographic rainfall

> **Examiner's tip** Focus on the processes, including lapse rates. Make sure your answer is better than GCSE standard!

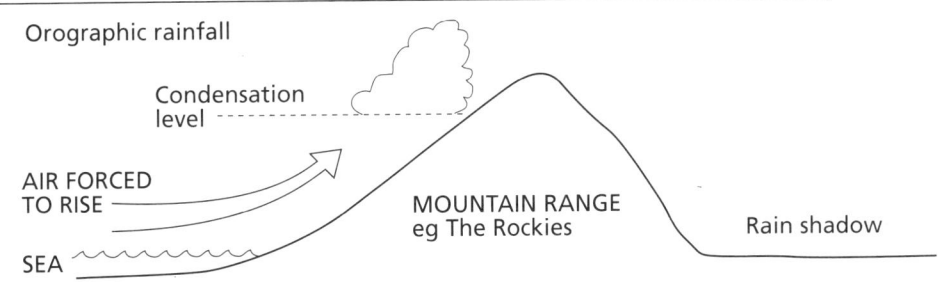

Orographic rainfall

[Diagram: Air forced to rise from sea over a mountain range (eg The Rockies), condensation level marked, cloud forms, rain shadow on leeward side]

Orographic precipitation is precipitation that results from the forced ascent of air over mountains. A mass of air, even if stable, will be forced to rise when it encounters a mountain range. When the air rises it loses heat as the mass of air expands. This type of cooling is known as adiabatic cooling. Winds that have blown over an ocean or a sea will have picked up large amounts of moisture and, if forced to rise, will cool. Cooling reduces the amount of water vapour the air can hold and condensation will occur when the dewpoint temperature is reached. Clouds will build up and precipitation will result.

The very high levels of rainfall on mountain ranges are enhanced by two other factors: first, in mid latitudes, depressions will bring rainfall, and the mountains will increase the degree of uplift and therefore precipitation. Secondly, the forced rising of air may produce conditional instability. While the rising air is cooling at the DALR, the cooling will occur faster than the ELR so the air remains stable, and will sink if it is able to, as it is denser than the surrounding air. However, once condensation occurs the air mass cools more slowly due to the release of latent heat. The air mass may then be warmer than the surrounding air and be unstable. The unstable air then rises and forms thicker, rain-bearing clouds, thus increasing rainfall. This is seen in the diagram below.

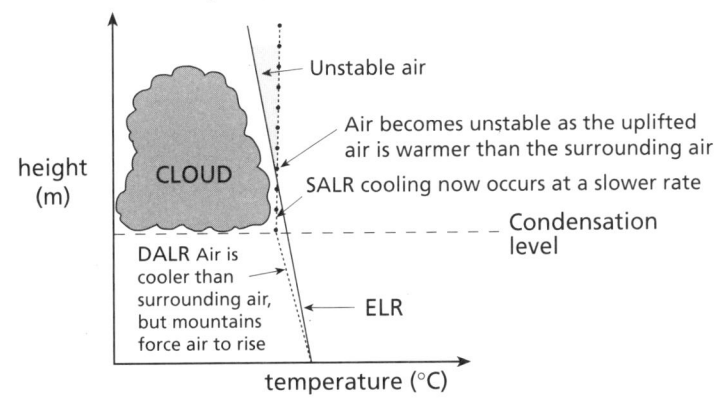

[Diagram: height (m) vs temperature (°C) showing DALR (Air is cooler than surrounding air, but mountains force air to rise), ELR, Condensation level, SALR cooling now occurs at a slower rate, Air becomes unstable as the uplifted air is warmer than the surrounding air, Unstable air, Cloud]

Conditional instability caused by orographic uplift

Answers to Unit 1

| Question | Answer | Mark |

Examples of areas with high orographic rainfall in Britain are the Lake District and Snowdonia where rainfall in excess of 2000 mm occurs, whereas lower lying valleys have far less precipitation. In the United States the Rocky Mountains have similar patterns of rainfall. (8)

Examiner's tip Inclusion of a diagram showing conditional instability is essential to obtain full marks.

| Question | Examiner's tip |

3
- Look at definitions given earlier for DALR and SALR. These terms allow meteorologists to forecast the likelihood of precipitation.
- **A** is a cold front. **B** is an anticyclone.
- Southern Britain is affected by anticyclonic conditions: the air is warm as it has its origin over the continent during summer. Lack of cloud has led to fog formation due to unimpeded radiation and cooling at night. The cold front has passed over the north of Britain so the temperatures are lower as the land is influenced by arctic maritime air and precipitation is caused by uplift of the warmer air at the front.
- Mists and fogs occur when the dewpoint temperature is reached. (See earlier answers for more detail.) These are made more frequent and persistent due to particles in the air.

4
- Tropical cyclones cannot form between 8°N and 15°S due to the lack of sufficient Coriolis force, but can only form over warm seas (+26°C) as moisture is needed to maintain the rising air currents by the release of latent heat. This is why they are unable to travel far inland or into higher, cooler latitudes.
- High winds, flooding and ocean storm surges.
- **A** (Florida) is in the developed world, so early warning systems will be in place allowing evacuation and amelioration of the impacts of the cyclone, building construction standards should be high and the economy is able to fund relief efforts. In contrast, **B** (Bangladesh) is poor, densely populated, very low lying and has a largely agricultural economy which will be harmed by cyclones.

5
- Define micro-climate, and explain the variables to be investigated – wind characteristics, temperature (ground and air), and humidity.
- Temperature. Albedo should be discussed, especially the diurnal variation between different vegetation types, (there will also be seasonal differences in deciduous woods compared to coniferous woods). Shade will be significant within woodland. Try to refer to fieldwork studies.
- Wind characteristics are changed by tall vegetation.
- Transpiration has an impact on humidity.

Answers to Unit 1

Question	Examiner's tip

6
- This question requires you to discuss urban microclimates, in other words the modification of weather systems (depressions and anticyclones) by built up areas.
- Urban areas are warmer, especially in the summer due to the different albedo characteristics and the lower specific heat capacity of concrete and tarmacadam.
- The incidence of summer thunderstorms is more frequent over urban areas due to convectional heating.
- Fogs and smogs are more common (Californian cities, for example Los Angeles).
- Winds are modified by buildings and architects now take account of this in design.

7
Collecting data on temperature, humidity, air pressure, precipitation, cloud types and wind characteristics in school would reinforce the understanding of the impact that anticyclones and depressions and their associated air masses have on the weather experienced. This information could then act as a control to monitor the impact that buildings and the school environment have on the microclimate, (see earlier answers for more information). If the school collected this information over a long period of time (20 years) it could then be used to examine the climate of the local area, and possible climatic change.

8
- Insolation (short-wave solar radiation) is reduced on entering the earth's atmosphere by absorption (by ozone, carbon dioxide, dust particles, and water vapour). Losses also occur through reflection by clouds, especially thick cumulonimbus clouds, and scattering.
- (b) is straightforward recall of basic information, clear annotated diagrams may help present your information clearly, give examples where necessary.
- There is a positive heat balance in low latitudes (40°N - 35°S), but a negative heat balance at the poles, if these areas are not to increase/decrease in temperature respectively there must be a heat transfer. Heat is transferred by horizontal and vertical heat transfers. Definitely use a diagram to help explain this section.

9
A lovely question! BUT it requires careful planning and structuring. Rain is condensed water vapour, two theories of water droplets exist - the Bergeron-Findeisen (ice crystal mechanism) and the collision and coalescence method (Longmuir). You need to explain under what conditions air picks up moisture and what mechanisms there are for uplift and cooling. You need to mention convergence (e.g. at the Polar front and the ITCZ), orographic and convectional uplift. Include diagrams.

Answers to Unit 2

2 THE BIOSPHERE AND ECOSYSTEMS

| Question | Answer | Mark |

1

> **Examiner's tip** This is a 'short answer question': keep to the lines on the booklet provided in the exam. Answer 2 only. It pays to learn definitions.

(a) Zonal soil: this is a soil in which climate and biota (especially vegetation) have been the major influence in determining the soil's profile. An example is a ferruginous soil.

Intrazonal: this is a soil where a factor other than climate determines the soil type, for example, parent material e.g. a rendzina (a calcimorphic soil).

Azonal: not had time to adjust. (6)

> **Examiner's tip** 5 marks, but not many lines. Think of the of the characteristics before starting to write. You have to use the same soil in (c), make sure it is suitable!

(b) A podsol. These have clearly defined horizons, and are fairly shallow. The litter layer is composed of poorly decomposed material. The thin mor humus layer is acidic. The upper part of the 'A' horizon is stained brown, but the lower part is eluviated and ashy in colour. The 'B' horizon is illuviated and has a reddish colour (redeposited iron and some humus). The 'C' horizon may be stony, and the parent material is often some kind of sandstone. (5)

(c) Podsols are of limited use as they are shallow, often stony, and acidic. Generally farmers use them for rough grazing or, increasingly, for forestry. The acidity and fertility may be reduced by liming and applying fertiliser. (4)

| Question | Examiner's tip |

2
- Describe the distribution of the taiga. Tree growth is limited to south of the line where no month has a temperature higher than 10°C. Adaptations to harsh environment include waxy needles, bark, sloping branches. You must include sketch diagrams.
- Podsols are mentioned in question 1. Label the diagram fully.
- High productivity of soft wood for pulp and timber, tolerant of wide range of conditions and infertile podsols. Forests are managed by replanting.
- Burning of fossil fuels results in acid rain. Acidification of the soil and water courses has caused death of trees. Some attempts have been made to counter acidity by application of lime.

3
- Another example of recall of information. Give brief definitions to the terms in (a).
- High biomass productivity in hot, wet tropical climates and rapid breakdown and recycling. The breakdown is slow in acid, cold conditions of taiga. This influences the humic content of the soil and influences processes like chelation.

Answers to Unit 2

Question	Examiner's tip
	• Note, the question says 'global effects': it calls for a discussion on the atmospheric consequences of loss of biodiversity.
4	• Straightforward question; recall of terms, and testing of how climate influences weathering and decay of litter. • (f) The removal of cattle for consumption leads to a decline in nutrients in the ecosystem and leaves the soil vulnerable to erosion.
5	Climax vegetation is in equilibrium with its environment, so time and environmental stability are key elements. In (b) mention plagioclimaxes and biotic climax: heather moorlands in Scotland and the Chalk Downlands are good examples.
6	• The material in questions 3 and 4 contains useful information to answer this question. Equatorial rainforests and Taiga *or* Tundra ecosystems are good choices. • Choose one only. People interfere in the ecosystems to increase the flow of nutrients to themselves (agriculture/forestry), for coastal protection or leisure. Mention changes to soil, vegetation and fauna.
7	• A sere is the sequence of changes that a community of plants within an area goes through as it evolves towards climatic climax. You need to select a xerosere e.g. sand dunes (psammosere) *or* a rock surface (lithosere) and to contrast with this a hydrosere, either a salt marsh (halosere) or pond infilling. • Suitable biomes to choose are savanna or rain forests. Discuss human activities (logging, grazing etc) but also natural changes e.g. climate.
8	Plan this carefully, presenting a balanced point of view, review the **evidence**, and give examples. There is good evidence from satellite photographs to suggest that the desert margin reponds to the amount of rain that falls each year, and you should include an account of the *variability* of rainfall in the Sahel savanna. Poor agricultural practices, such as overgrazing and destruction of woodland for fuel have also contributed to the problem. Discuss ideas of carrying capacity for these environments. Include positive actions by people to reverse desertification such as tree planting.

Answers to Unit 3

3 HYDROLOGY AND DRAINAGE BASINS

| Question | Answer | Mark |

1

> **Examiner's tip** You could achieve full marks by using the formulae, or by using words. To define hydraulic radius you could use a sketch.

(a) (i) Discharge is the volume of water passing a point. It is calculated from its cross-sectional area and velocity. It is expressed as m^3/sec^{-1}.

$$Q = A \times v \qquad (2)$$

Hydraulic radius is a measure of channel efficiency. It is the ratio between cross-sectional area and the wetted perimeter (Pw), where the wetted perimeter is the boundary between the water and the channel.

$$R = A / Pw \qquad (2)$$

(ii) These are very important in determining the velocity of a stream. Velocity is determined by the the gradient, so the highest velocities might be expected where the river is flowing down a steep slope. However, there are very high frictional losses in upland streams due to the irregular channel. Hydraulic radius is a measure of the losses of energy that a channel experiences due to friction, and Manning's (n) measures the effect of channel roughness on velocity. Other factors influencing the velocity of a stream are the sinuosity of the course and the amount of vegetation found within the channel. (3)

> **Examiner's tip** Calculation of lag would gain 1 mark.

(b) (i) The lag time of a flood hydrograph is the length of time that has elapsed between the precipitation peak and the peak storm discharge of the stream. (2)

Highest rainfall 3 hours, peak discharge at 10 hours, so lag is 7 hours. (1)

> **Examiner's tip** The wording of the question indicates that a discussion of the relative contributions of base flow versus storm runoff is the main point here.

(ii) Base flow is water that comes from ground water or aquifers and is controlled by seasonal factors rather than individual storm events. This is because water flows very slowly at depth. Runoff, however, is much more influenced by individual precipitation events. For all soils there is a limit as to how much water can infiltrate within a given time, and the excess will contribute to run-off. In this case the rainfall was intense and heavy (45 mm rain fell in 6 hours), so much of this ended up in the river. (5)

> **Examiner's tip** To earn the maximum marks here you would need a detailed account including examples.

(c) Rivers flood when the ability of the channel to transport water downstream is exceeded, i.e. bankfull is exceeded. There are various factors that contribute to this: impermeable rocks, little vegetation to intercept rainfall, saturated soils and high tides that 'pond up' water and prevent it from draining to the sea. High precipitation or meltwater in spring may result in flooding. Monitoring a river's

Answers to Unit 3

Question	Answer	Mark

discharge, particularly previous flooding in response to specific events, will enable prediction of future events, given the characteristics of the catchment area. It is possible to plot the discharges of a certain size against the frequency of occurrence and this allows scientists to predict the recurrence of a flood of a certain size, and hence the need for flood protection measures can be assessed.

The effects of floods can be reduced by a whole range of measures and they are generally divided into flood protection methods, which aim to protect the areas under threat, and flood abatement, which aims to reduce the water entering the river at a particular time by reducing 'flash flooding' and reducing the storm peak. Flood abatement is directed at the whole drainage basin, especially the upstream area, and includes such measures as planting trees in the catchment areas, and preserving areas of swamp (perhaps as nature reserves to act as 'giant sponges' within the drainage basins), and reducing run-off from farm land by contour ploughing and terracing. Flood protection measures include straightening sections of channel, building dams (for example, the Clewedog dam in Powys), and constructing artificial levées to allow rivers to carry more water (for example, the Mississippi). (10)

Question	Examiner's tip

2
- Part (a) relies on recalling definitions. (b) is easier than it might appear, you have to draw appropriate *idealised* **cross-sections**. Note the braiding in photograph B. Label with words such as turbulent, variable discharge.

Cross-section of channel A. (Mountainous upland stream)

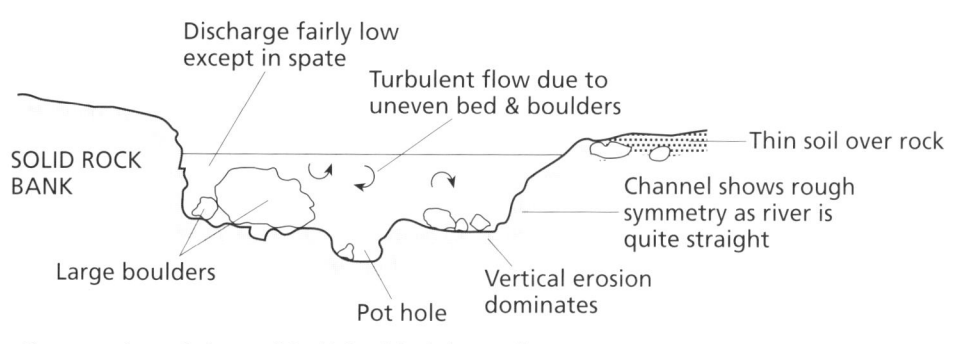

- Discharge fairly low except in spate
- Turbulent flow due to uneven bed & boulders
- SOLID ROCK BANK
- Thin soil over rock
- Channel shows rough symmetry as river is quite straight
- Large boulders
- Pot hole
- Vertical erosion dominates

Cross-section of channel B. (A braided channel)

- Banks of unconsolidated sands and gravel
- Coarse debris deposited as mobile bars during low flow
- Unstable banks and bars eroded
- Discharge very variable eg proglacial or periglacial environments
- River splits into several channels (anastomosing)
- Fine sand deposited in slow flow
- Flow may be very turbulent

- Draw labelled sections in (c). Consider the differences in lowland/upland valleys. Other factors include human activity (such as bank protection), rock

Answers to Unit 3

Question	Examiner's tip
	type (contrast limestone gorges with a valley formed in less resistant rock), past erosional history e.g. glaciation.
3	(a) Use interpolation. Make sure you refer to river processes in (b). In (c) give 3 points for each part. In (d) consider the changes to the efficiency of the channel (friction) and hence the river velocity, and consequent changes to erosion and deposition.
4	Look at the answer to question 1. Plan this essay carefully, making sure your ideas flow in a logical sequence. Include a wide range of points and detailed examples to back up your points. Droughts may be caused by climatic factors, but people now extract large amounts of water and have changed the run-off characteristics of drainage basins. Examples, from California or the Indian subcontinent may be useful, especially schemes to ameliorate the impact of floods/drought.
5	Make sure you have discussed drainage patterns (i.e. dendritic, trellis, radial). Cover factors that operate currently (e.g. geology, structure) as well as antecedent, superimposed and patterns modified by deglaciation (river capture) and tectonic movement. Examples may include Pickering (North Yorkshire), the Lake District, scarp and vale in SE England.

4 ARID AND SEMI-ARID ENVIRONMENTS

Question	Answer	Mark
1		

Examiner's tip Learn definitions, they often crop up on the exam paper. Only define two; three are defined here so you can check your answers!

(a) Exogenous: these are permanent rivers which have their source in a non-arid region and only flow through a desert on the way to a sea e.g. The Nile.

Endoreic: these are rivers which terminate in a lake within a desert area.

Ephemeral: these are temporary streams which flow after heavy rainfall. (2 + 2)

Examiner's tip Candidates too often fail to read the instructions, it says '*Describe*' in (i), and in (ii) '*account*' for **one**.

(b) (i) The features are wadis (arroyos), which are steep sided ravines, and alluvial cones, which are fans of coarse river debris. Where the latter merge, bahada (a continuous sand and gravel deposit) is found, and playas are level areas covered in fine silt occupied by lakes after rain. (2 + 2)

Answers to Unit 4

Question	Answer	Mark

    (ii) An alluvial cone forms when flash floods carrying heavy loads leave the confined space of a wadi. The stream breaks into distributaries and the decline in competence and capacity results in deposition. Repeated events will build up a cone. (4)

(c) Desert people are often nomadic and move to areas where there has been rainfall to exploit the sudden growth in vegetation that follows rainfall in the desert. (3)

Question	Examiner's tip

2
- The dunes at *X* formed by aeolian processes, sand sized material is transported by saltation and surface creep. Draw diagrams to show processes. The lack of sand has produced the characteristic barchan dune form to the middle right of the photograph.
- The surface at *Y* is a hamada or reg, these are 'stone pavements'. Here the surface is covered in stones and boulders (hamada) or fine material (reg) derived from weathering from which finer material is removed by deflation. Sheet flooding may also remove some of the fine material.
- At **Z** the retreat of the mountain front has formed a pediment surface with associated bajada (bahada). These are mentioned in question 1. Mention King's pediplanation hypothesis.
- In areas of the Middle East dune movement is reduced by replanting, bulldozing and spraying with oil, for example near roads. To prevent the covering of fields in semi arid areas suffering desertification, planting is used.

3
- The features shown are dealt with in questions 1 and 2. The playa is a lake bed covered in fine clays after rain: it may be filled by a temporary lake.
- During pluvials, the climate was much wetter than at present. Many features such as wadis and those mentioned in 1(b) are very impressive features, yet the processes observed today are not regarded as being capable of their formation. There is a mass of historical and archaeological evidence to suggest that people grew crops in areas that are semi-arid today. Evidence includes that from lake terraces and fossil pollen.

4
- Draw annotated soil profiles. Mention solonchaks, and solonetz. The soils are highly alkaline, white/grey in colour, capillary action leads to salt accumulation (+0.3% NaCl and $NaSO_4$) and a very low humic input. You must mention soil processes.
- In (b) refer to case study material.

5
- Flash flooding, explain why rain should occur. See answer to question 1.
- See answer to question 3. Water formed features especially are thought to result from pluvial periods, though their formation continues today.

Answers to Unit 5

5 COASTAL ENVIRONMENTS

Question	Answer	Mark

1

> **Examiner's tip** This is a very straightforward question which allows you to recall basic information and apply it to a new situation. Make sure you only give the information asked for. (a) (ii) is easier to answer with annotated diagrams, and is more likely to obtain full marks.

(a) (i) Landform A is a compound spit. (2)

(ii) Spits are elongated, narrow accumulations of shingle and sand which join the land at the proximal end. Material is moved along the coastline by longshore drift (LSD) and where the coastline changes direction, for example at an estuary, the debris will build out into the sea. The distal end has several 'hooks' known as laterals. These form due to the effects of wave refraction or waves from a secondary dominant wind direction. Each recurve represents a former distal end of the spit, so the successive growth over time can be seen. (6)

> **Examiner's tip** Make sure you make three points.

(b) (i) Area B is a salt marsh which is a low lying accumulation of mud, much of which is covered at high tide. The area is covered by halophytic vegetation, and is dissected by creeks. (3)

(ii) The vegetation in salt marshes is halophytic (salt loving), as the low marsh is inundated twice a day by tides and only plants with special adaptations can survive in these conditions. The marsh shows a succession of vegetation types known as a halosere. The lowest areas are covered for most of the time, and have the most salt tolerant plants like samphire. Higher up, plants like sea asters can grow. (5)

> **Examiner's tip** It is important to make the connection between the change in vegetation, (the halophytic succession) and accretion. Good candidates will not forget human influence on vegetation and would gain extra marks. Full marks would require reference to case study material.

(iii) Salt marshes form in low energy environments by the accumulation of mud on which plants become established. Mud closest to the sea is covered for most of the tidal cycle, only 'pioneer' species tolerant of saline conditions can establish here, and these help trap and stabilise the mud. The area accretes, so it is covered by the sea for less time, and sea lavender and grasses establish in the zone covered by spring high tides. Eventually, the marsh reaches a height where sea water no longer covers it and non-halophytic plants enter the succession; the area will then either evolve towards the climax vegetation or will be reclaimed for agriculture producing a plagioclimax. (5)

Answers to Unit 5

| Question | Answer | Mark |

2

> **Examiner's tip** Spend several minutes studying the maps, aiming to identify the types of landforms present, such as cliffs, WCP and bays. You have been given a geology map, so geology must have a major impact on the coastline. Try and relate features to lithology and structure. One of the rocks is millstone grit, not known as an easily eroded rock - why does it form a bay? The bay must be influenced by the syncline. Plan your essay, then write an introduction and refer to controlling processes.

Coastlines evolve their characteristic landforms by the interaction of marine processes, subaerial weathering and mass movement and these are influenced by the lithology and geological structure of the area. The coastline shown on the map extract is a cliffed coastline with small bays and headlands, as well as quite extensive wave-cut platforms. Deposition has produced small bay head beaches.

> **Examiner's tip** Weak candidates often adopt a 'guided tour of the coastline' approach; it is far better to identify features and processes. This avoids repetition.

Most of the coastline has well developed cliffs: limestone is a strong rock and can support steep rocky cliffs. The cliffs are higher on the headlands and the reason for this is explained later. The cliffs appear to be actively undercut because of their steep profile and the position of high tide, however the upper slopes are more gentle suggesting that mass wasting might be important here.

Cliff profiles on the south coast of the Gower Peninsula

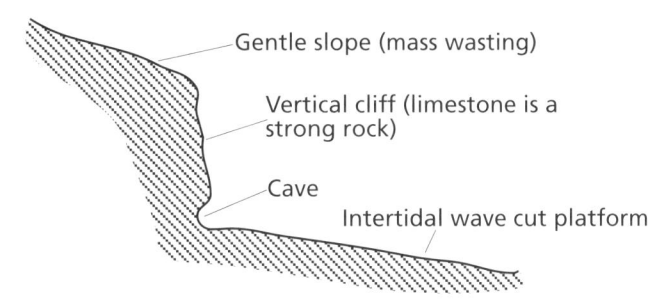

- Gentle slope (mass wasting)
- Vertical cliff (limestone is a strong rock)
- Cave
- Intertidal wave cut platform

The retreat of the cliffs has left a prominent wave-cut platform. As this rocky shelf is exposed at low tide, but covered at high tide, it must be what Bird identified as an inter-tidal platform.

There are limestone outcrops on the southern part of the coastline, but the resistance of the rock varies, producing the fretted nature of shoreline. Some of the bays are fault controlled, for example, the geology map shows a fault cutting the bay to the east of The Knave (435863). Faults are weak points as movement has occurred along them crushing the rock. The sea is a very selective agent of erosion and readily exploits these weaknesses by corrasion, hydraulic action and solution. Any lithological variation or variation in the joint density will also be picked out, and at GR 437858 there is a cave marked.

> **Examiner's tip** Able candidates will be able to link the theoretical aspects of coastal development to the map, so in the case of headlands you need to discuss the geological controls on their formation and wave refraction. Some credit would be given for recognising that headlands and bays result from discordant rocks.

Answers to Unit 5

Question	Answer	Mark

Port-Eynon Point and Oxwich Point are headlands, the rocks here are being eroded less rapidly than the millstone grit in the bay. The grit is a resistant rock and is younger than the limestone, however it is only preserved in the syncline and here the rocks are lower in height, allowing the sea to erode them more easily. Low cliffs will be eroded faster than high cliffs as a greater proportion of the cliff is exposed to attack from the sea between high and low tides.

> **Examiner's tip** The aim throughout any essay is to produce a logical development of ideas and discussion, because this is one of the things examiners look for in awarding 'A' grades. An example of the 'flow of ideas' should be apparent in the last paragraph where wave refraction acted as a link between the coastal erosion features and the deposition in the bay.

Erosion will be concentrated on the headland by wave refraction, as the sea shallows further out, causing the wave crests to 'wrap around' the headland. In the bay the energy of the wave is reduced because it is spread over a wider zone and waves with a lower height are likely. Deposition results in the lower energy environment, and the bay has a 'classic' swash aligned beach (crescentic shape). Erosion of the cliffs provides a source of sediment. Sorting of the beach materials is apparent with sand found at the north-eastern end and coarser material at the south-western end of the beach, this may result from drift of material along the beach. Sand, being lighter, will be moved preferentially and the heavier material will remain in situ until the processes of attrition wear it down.

The formation of erosional and depositional features along this stretch of coastline may be seen as a result of a number of different processes, and influenced by the geology and structure of the area. (23)

Question	Examiner's tip

3
- Part (a) is straightforward. In (b) you are able to discuss all features found on cliffed coastlines as the question does not limit you to cliff 'profiles', but they must be erosional ones. Include wave cut platforms, caves, stacks, arches.
- Select examples of places where spits and marshes have changed the coastline. e.g. Blakeney, North Norfolk since medieval times, Spurn Head spit on the Holderness coast, or cuspate foreland on south coast at Dungeness. Draw sequences of labelled diagrams.

4
- Your answer should be confined to spits, tombolas and cuspate forelands, as features must result from LSD.
- Discuss beach protection methods such as groynes and stabilising sand dunes on spits. Put these measures in the context of the sediment cell. Erosion may result in down drift once the sand supply is disrupted.

5
- Two terms need explaining, first, structure; this refers to the relationships of the rock bodies to each other, i.e. the bedding relationships, the juxaposition of one rock type to another and features such as bedding, jointing, faults and

Answers to Unit 6

Question	Examiner's tip

folds. Second, morphology of the coast means the shape of the coastline and its features. Coastal erosion is a selective process exploiting any weaknesses in lithology (rock type) and structure. Evidence for this can be seen at the macro scale in concordant and discordant coastlines; examples such as Lulworth Cove illustrate this well. Features such as stacks and arches often reflect structural weakness, and at a micro scale lithological differences are picked out in the cliff profiles and wave cut platforms: this can be seen very clearly along the coastline of North Yorkshire in Robin Hood's Bay, but try to quote examples from your own field work.

- The morphology of coastlines which are a function of coastal deposition reflects the amount and type of sediment available, the shallowness of the coast and the processes that act, e.g. aeolian/LSD, type of waves and even vegetation.

- The statement in the question is generally true, certainly for coastlines dominated by erosion. Remember the height of the land is often influenced by geology (the south-east of Britain is made of less resistant and indurated rocks), but some processes will override this tendency, for example features formed by sea level change.

6
- Explain corrasion (abrasion), corrosion, hydraulic action, quarrying, but not attrition, which refers to the erosion of the beach material.

- Include isostatic and eustatic sea changes. In Scotland there are raised beaches, and fjords, and on the coast of Ireland and S.W. England, rias. Other features are fjards, Dalmatian and Atlantic coastlines. To determine the second part of the question consider what processes would operate on the 'new' coastline, the depth of water and materials available for deposition.

6 GLACIAL AND PERIGLACIAL ENVIRONMENTS

Question	Answer	Mark

1

Examiner's tip This relies on recall. If you have not revised, you will lose easy marks.

(a) The tundra climate has mean average temperatures of 3 degrees C or less. The temperatures are low all year round, with 6 - 10 months below freezing. Tundra climates show considerable variability depending on their proximity to the sea, and a range of winter temperatures from -10 degrees to -40 degrees is experienced. Precipitation is low at 250 - 300 mm, but may be higher in maritime areas. (4)

Examiner's tip This is a straightforward question, so examiners will be looking for accurate descriptions, of a high standard, of features and the processes responsible for their formation. This should include alternative theories of formation where appropriate. A set of diagrams is the best way of tackling this question.

Answers to Unit 6

| Question | Answer | Mark |

(b) (i) **Patterned ground**

Patterned ground is the collective term for the arrangement of different sized stones and sand into geometric patterns such as polygons, and can be regarded as the 'surface decoration of the regolith'. On almost-level surfaces there are polygons, with the coarsest material forming the outsides, and finer material inside. On steeper slopes these shapes become elongated into 'stone garlands', on slopes 7-10 degrees they become stripes, and on very steep slopes they disappear.

Sorted stone polygons and stripes result from a series of processes. First, the production of extensive block fields ('felsenmeer') by intensive freeze-thaw action, and then sorting of material. The sorting process relies on cryoturbation (the repeated freezing and thawing of the active layer).

The processes forming patterned ground are not clear, but once the pattern is established, the method by which it is perpetuated is clear. Coarse-sized material is porous, but finer sediment can hold more soil moisture. Areas with larger stones will freeze first because of their greater thermal conductivity. When the water in the fine sediments freezes, a mound will be produced, and any larger stones will roll down the sides into the gaps between the mounds, leaving only the finer sediment behind. The formation of polygons is an example of positive feedback as this process of freezing and thawing will lead to the differentiation of the sediment by size, and the finer sediment is associated with higher soil moisture which produces the mounds.

The coldest areas have larger patterned ground features, called ice wedge polygons, formed by thermal contraction. Intense freezing in winter causes cracks within the permafrost and active layer (rather like desiccation cracks on mud puddles). These fill with water during the summer thaw and later freeze. The gradual development of large polygonal ice wedges produces a network of polygonal sediments with lakes in the centre. (6)

(ii) **Pingoes**

These are domed shaped hills with a perennial core of ice. They may be 50 - 60 m high, and have a diameter of about 500 m. They occur in areas of low topography, usually in sandy sediment, and some have a central crater with a pond.

Two theories of pingo formation are recognised: first, the Mackenzie type, or closed system, and second the East Greenland type, or open system. These are shown in the sequence of diagrams below and overleaf. In the open system water continues to flow into the talik.

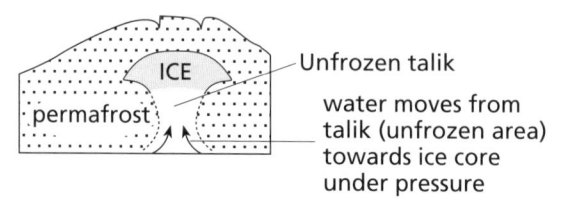

1. Open system (East Greenland type)

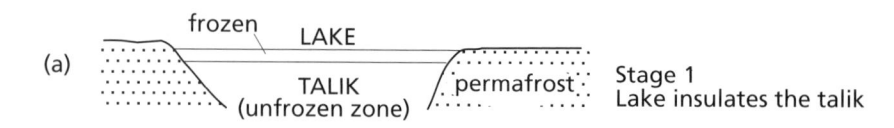

(a) Stage 1 Lake insulates the talik

Answers to Unit 6

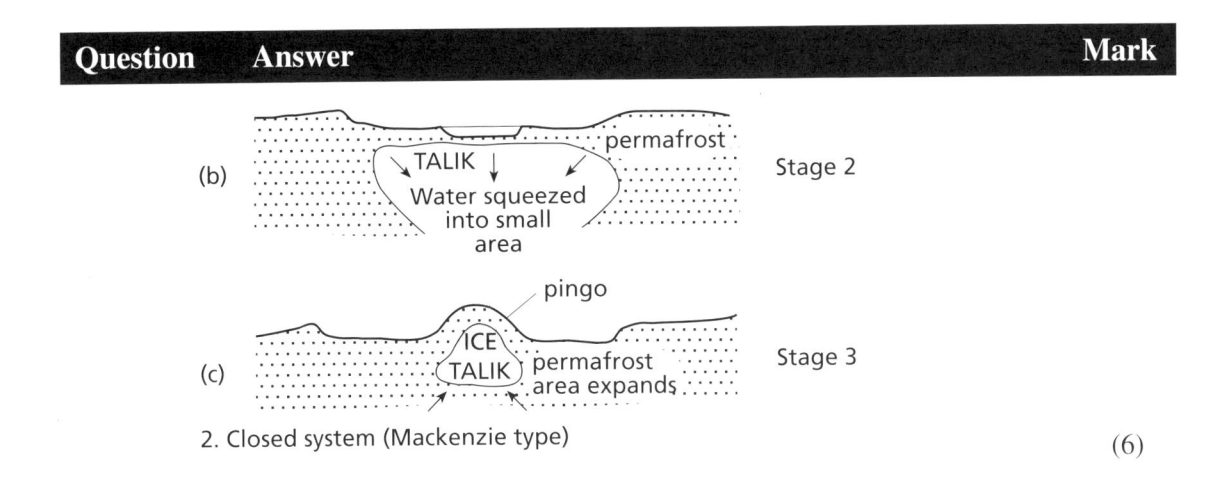

Question	Answer	Mark

(b) *[Diagram: Stage 2 — permafrost with TALIK, water squeezed into small area]*

(c) *[Diagram: Stage 3 — pingo with ICE/TALIK, permafrost area expands]*

2. Closed system (Mackenzie type) (6)

Examiner's tip It is important when tackling this type of question to be able to give examples of plants growing within these differing habitats. You will be familiar with some of these plant species if you have visited the Alps or upland areas of Britain. Be able to explain the physiological adaptations of plants and problems for plant growth in the tundra. Without this kind of detail you will not achieve the two highest grades.

(c) Lowland tundra areas, following deglaciation, would be characterised by a lithosere. A lithosere is the succession of plants that colonises dry rocky surfaces. The climax vegetation is dominated by herbaceous perennials, sedges, grasses and creeping shrubs like dwarf birch. The flora has few species, plants growing in these areas have to be highly adapted and many of the species are endemic.

Initially the areas will lack any type of soil development, organic matter and plant nutrients. Pioneer species in all habitats will be lichens and mosses as these can grow without soil. To begin with these are just encrusting lichens, but gradually dense growths of reindeer moss establish where the regolith is undisturbed.

Within small areas of the tundra there are considerable variations in vegetation caused by variations in the micro-habitats, and these have different plant communities. Important variations include aspect, shelter, substrate and availability of moisture. In poorly drained hollows and on the valley floors, or anywhere conditions are saturated, sphagnum mosses and cotton grass establish. These build up quite thick layers of partially decomposed material, and from a distance the bright green mosses and white heads of the cotton grass are a stark contrast to the grey expanses of moraine and outwash gravels. On better drained surfaces sedges and grasses capable of surviving periods of drought are found. In contrast, sunny, sheltered and well drained sites are colonised by cushion-like plants like the saxifrages and arctic poppies. In well drained sites with scree-like material, bilberry and crowberry plants are found, and in the least exposed areas with sufficient water, dwarf varieties of trees like birch grow. (9)

Answers to Unit 6

Question	Examiner's tip

2
- A risk with this question is wasting time on features *not* on the map (e.g. some glacial deposits). Give GR and explain why they are evidence for glaciation. What is there: corries (note orientation and shape), parabolic troughs, ribbon lakes and alluvial infilling, hanging valleys, fjords (Kinlochleven has an aluminium works, and the buoys GR 1361 might make you aware that it is a sea loch), and rocky surfaces suggest soil has been removed by glaciation.

3
- Explain, using sequences of diagrams, the following features: X a truncated spur (note the scree development at the base of the slope, this contributes to the parabolic shape of glacial troughs). Y is a corrie (cwm/cirque). Discuss nivation, orientation of corries (insolation), compressive flow at the lip and the contribution of debris by frost shattering. Surface Z was topographically above the glacier which filled and overdeepened the valley, it is not an 'Alp', as it is not the shoulder of the valley (the pre-glacial land surface). Two alternatives exist: it may have been snow covered and subject to periglacial processes or, like parts of Scotland, covered by ice and 'protected' from erosion (Sugden's research on Scottish glaciation). Make sure your answer is of a higher standard than GCSE!

4
- In warm (temperate) glaciers the ice at the base of the glacier is at or above the pressure melting point of ice, but in cold glaciers the ice at the base of the glacier is below the pressure melting point of the ice, so the ice is frozen to the bed rock and hence has zero velocity.

- The rapid flow of temperate glaciers allows erosion, whereas cold based glaciers are responsible for very low rates of erosion, due to the absence of sliding. Ground moraine is carried by temperate glaciers, but cannot be transported by cold glaciers. Fluvioglacial landforms will be present only in temperate glaciers.

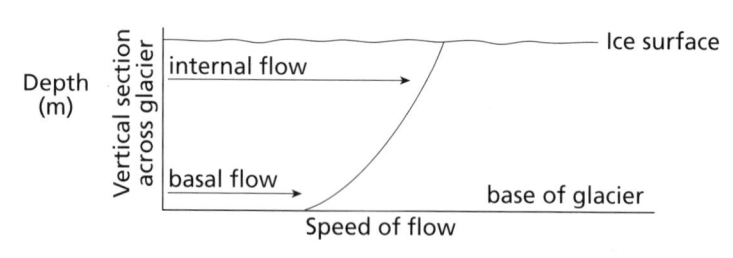

a. Changes in velocity of ice within a glacier

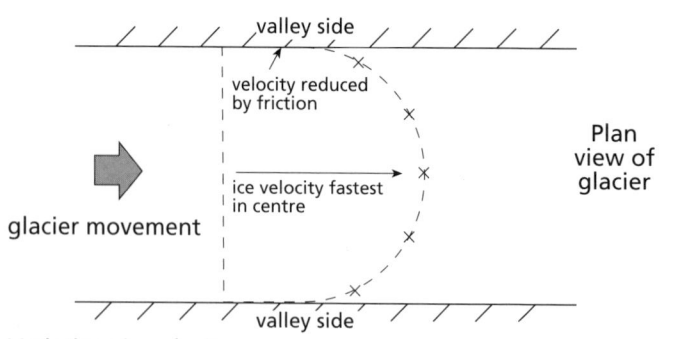

b. Changes in velocity of ice across the surface of a glacier

Variations in velocity across and within a 'warm' glacier

Answers to Unit 7

Question	Examiner's tip

- Glacial surges occur when ice is moved down the glacier at much faster speeds than normal, perhaps 10-100 times the normal velocity. This may cause a dramatic advance. The cause of surging is not clear, but is probably connected to an increase in snowfall in the glacier. The surges are often accompanied by increased meltwater and hence sliding.

5
- Describe both sub-glacial features like eskers and pro-glacial features like outwash plains. Some, like kames, form on the ice margins. Streams in periglacial areas are braided; they have extremely variable discharges as they are frozen most of the year, but get large amounts of meltwater in spring. This runs over the surface as the permafrost is an impermeable surface. The heavy, coarse load can only be carried by streams which have steep gradients.

    Some problems of development are in the revision summary. Positive aspects: glacial areas have meltwater for HEP and are appealing to tourists and skiers.

6
- (a) This is basic knowledge. (b) The key here is to recognise that deposits have different modes of formation and transportation. Fluvioglacial deposits will be coarsely sorted, rounding is present, but minimal, as transport is usually short, and fine clay sized material is washed out. Glacial deposits are mixtures of fine clay matrix with angular frost shattered and plucked boulders (deposits often have a bimodal distribution). Solifluction head deposits are recognised by the angularity of debris and orientation of the long axis in direction of movement. The short distance of transport means that the head is the same material as the underlying rock. There may be evidence of stripes/polygons/ice wedges/lobes.
- Deposits may be small and irregular compared to flood plain deposits.

7 THE LITHOSPHERE AND PLATE TECTONICS

Question	Answer	Mark

1

Examiner's tip This question is founded on basic factual material: if you have found it hard, revise more thoroughly, making sure you know the basic features and processes occurring at plate margins. The art of these questions is to answer in a concise form, including only the material asked for. Plan your sentences before you write.

(a) (i) Crustal creation occurs at this margin due to the upwelling of magma. (2)

(ii) Heat from the decay of radioactive isotopes produces convection currents within the mantle and these pull the plates apart. The plates are also subject to pull by subducting plate margins. (3)

Answers to Unit 7

Question	Answer	Mark

(iii) The evidence for spreading is: first, the pattern of magnetic stripes on either side of a ridge. Second, the pattern of heat flow, which declines from the spreading ridge, and third, the topography of the sea floor. (3)

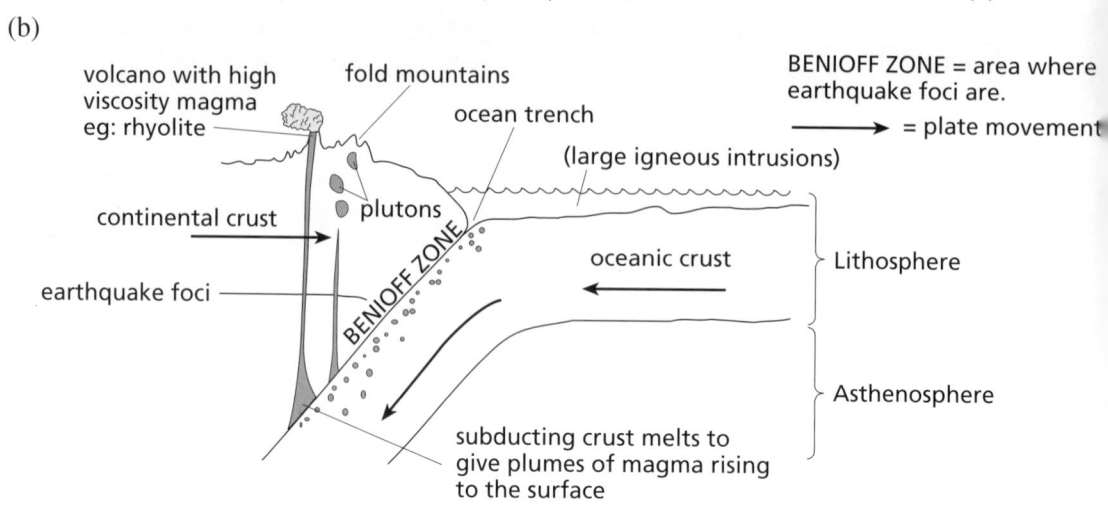

(b)

Diagram labelled: volcano with high viscosity magma eg: rhyolite; fold mountains; ocean trench; (large igneous intrusions); plutons; continental crust; BENIOFF ZONE; oceanic crust; earthquake foci; Lithosphere; Asthenosphere; subducting crust melts to give plumes of magma rising to the surface. BENIOFF ZONE = area where earthquake foci are. → = plate movement

The subduction and subsequent heating of the plate results in partial melting, and this magma rises due to its lower density. It may be intruded into the crust to cool as batholiths, or some may be erupted as andesitic volcanoes. The trench forms as the oceanic crust is being pulled down. Fold mountains are produced by the buckling of oceanic trench sediments. (8)

(c) Earthquakes result from the movement of the subducting plate. This movement does not occur constantly, due to friction. Friction holds the plate in place until the stresses are too great and the plate suddenly moves. The earthquake foci lie on the Benioff zone (an inclined plane). Deep focus earthquakes will have had their energy dissipated as they travel through the overlying rocks, but shallow focus quakes will have a greater intensity. (5)

(d) Volcanoes at constructive margins are effusive. At the constructive margins basalt, a 'runny' form of lava, is generated and erupted. The lava erupted at destructive margins is richer in silica, and is viscous ('sticky'). These volcanoes are steep sided since the lava is viscous and cannot flow very far. Pyroclastic and ash flows are more common at destructive margins. The eruptions are explosive as the gases are unable to escape readily. (4)

Question	Examiner's tip

2
- Three types of margin, diagrams — save words. Constructive: ridges with associated central rift valley. Normal faulting. Basalt erupted to create new oceanic crust, under water except volcanic islands. Destructive: at ocean/continental fold mountains andesitic lavas erupted from steep sided volcanoes, subsurface granitic batholiths, ocean trench, but at oceanic/oceanic margins get island arcs e.g. Tonga Islands, no fold mountains. Where continental collision occurs, get suturing of continental land masses and fold mountains from intervening oceanic sediments, e.g. Alps from closure of Tethys Ocean. Conservative: no crustal creation or destruction, may get some indication of boundary - rumpled hills, valleys, but

Answers to Unit 7

Question	Examiner's tip

often no indication. Visually, the least impressive plate margin. (The author has crossed points along the San Andreas boundary where on the ground there is nothing to be seen.) Evaluate your answer.

3
- Volcanic form, is a reflection of vent shape (vent/fissure) and lava type. Runny basalt lavas give effusive eruptions and gentle volcano shapes. As lavas become more viscous, the eruptions include more pyroclastics (ash and solid particles), composite cones may form. The most viscous lavas formed at destructive boundaries are associated with explosive eruptions, rhyolite spines and pyroclastics. Distribution reflects boundaries, explain lava genesis. Use diagrams to illustrate your points.

- Lava flows usually pose a low threat to life, and diversion from property is possible by excavating channels to lead the lava away from settlements. The big killers are lahars and nuées ardentes, especially as these explosive eruptions have proved more difficult to predict. Prediction allows evacuation, e.g. Mt Pinatubu.

4
- Plate boundaries. Sea floor spreading. Draw neat diagrams. Normal faulting is produced by the extensional environment, and get a rift valley in centre of the ridge. In (d) suitable hazards would be vulcanism or earthquakes.

5
- Look at diagrams 3A and 3B: it is an oceanic/oceanic destructive margin. Describe and explain. Note how the earthquake zone is extensive in Map A due to deep focus quakes. Risks from tsunamis include deep water close to the shore and building near coasts. Think about tourism/fishing.

- The classic example of a hot spot island chain is Hawaii, also Bowie Island off Alaska. The anomalous hot area of the mantle generates lava which melts through the crust and builds up a volcanic island; as the plate moves, another volcanic island forms, producing a chain.

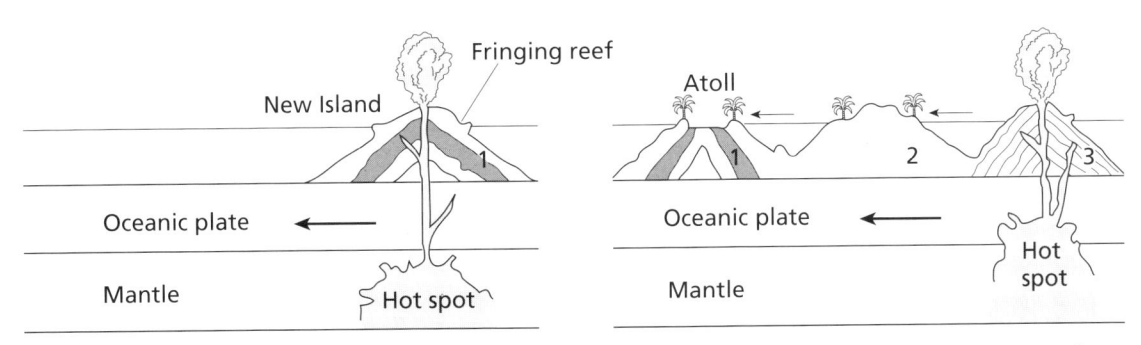

A volcanic island is built up by extrusion from a stationary hot spot in the mantle.

As the plate moves the volcano (1) is carried further from the hot spot and becomes extinct. A new island forms over the hot spot. Continued movement forms a chain of islands whose age increases away from the hot spot.

Formation of a chain of islands by hot spot activity

Answers to Unit 8

8 SETTLEMENT

| Question | Answer | Mark |

1 (a) (i) The old walled city (original site) is on a bridging point and has expanded to the west. On the east bank, the CBD is found. It is here because it was probably established by a colonial power, after the walled city, hence it is outside it. It is still relatively central, as at the time it was built, the city would have been much smaller. Directly to the east, is the colonial area. It is here because of its proximity to the CBD (access) and it is away from the old walled city.

> **Examiner's tip** The candidate has combined both descriptions and reasons as the question demands. Weaker candidates often only give descriptions.

Newer middle and high class housing has located adjacent to the colonial area, explained by Burgess'/Hoyt's 'attract and repel' idea. Like land use (i.e. high quality housing) is attracted to like. Also, these suburbs are repelled from lower income and shanty town developments because they are undesirable.

The old industrial area is located near good communications (rail and road), so it is accessible. This area is bordered by the new low class housing, so the residents can be near work, because they can neither afford to travel, nor escape industrial pollution. The newer industrial area is also next to the low class housing, on a main road (modern transport needs). It is further out because there is little room for it centrally, and land is cheaper and more is available for larger sites. Finally, shanty towns are found on unwanted land e.g. marsh that is prone to flooding. Some is near the CBD, probably built on waste ground. Although shanty towns are last in the pecking order (Bid Rent Theory), they will locate near industry and the CBD, to be near to possible jobs. (5)

> **Examiner's tip** By applying model theories, locations can be explained. This is a very thorough answer, taking each land use in turn.

(ii) Models are not descriptions of specific places, but generalisations. Therefore, no model can fit all cities exactly.

> **Examiner's tip** A vital point, well described. Models are predictive, not descriptive.

If the model is compared with Port Harcourt (Nigeria), there are similarities. New industry, such as the Shell/BP complex, is found on the city outskirts, near road and rail lines. Also, many of the shanty towns are found on unwanted sites near the Amad Creek. However, there is no walled city in Port Harcourt, because the city dates back primarily to 1916 when the British colonial railway reached it from the Enugu coalfields. This is why the settlement is sited on a navigable river (transport), perhaps a similar reason to that in the model.

> **Examiner's tip** It is perfectly acceptable to use an example that does not have all the characteristics of the model, but it is best to explain any variations.

Finally, the model acknowledges the constraints of physical barriers. Port Harcourt is similarly restricted by the Benue and Bunny Rivers. (5)

Answers to Unit 8

| Question | Answer | Mark |

(b) The population pyramid shows an 'unnatural' bulge in the 20-45 age group for men. This is because, in many regions of the Developing World, particularly parts of Africa and Asia, the majority of migrants moving into cities are young, able-bodied men searching for work. An absence of women may reflect culture and patriarchal societies. In much of Africa, target migration (e.g. saving enough money to marry) often explains this pattern. In the other age groups there is parity between the sexes, as few migrants in these groups arrive to distort the natural pattern. (5)

(c) (i) The basic problem that Mexico City experiences is being a victim of its own success. The population has rapidly increased from 1.7 in 1940 to c.18 million in 1990, and is estimated to reach 25 million by 2000 AD. Even though migration has slowed and natural increase accounts for a greater percentage of growth, 1000 migrants still arrive each day. Thus, the city is over populated, i.e. there are not enough resources for the number of people and those resources that are available are unevenly distributed. The urban density of the 'Ant Heap', as it is locally known, is very high (12 000 pop./km^2): on average this is three times the urban density of London.

This population explosion leads to problems in housing provision. One third of families live in only one room, whilst half the houses have been illegally built. This unregulated growth has led to problems with sanitation and disease.

Examiner's tip 'Problems of sanitation and disease' is a very general statement. Are there any facts to illustrate this?

Private rented 'vecindad' housing is little better. One in ten homes do not have electricity; a similar percentage to those who have no water supply. There is a housing shortage: temporary shelters from the 1984 earthquake are still occupied. Other services, such as education and health are inadequate and unevenly distributed, but Mexico City is still better off than anywhere else in Mexico, so it continues to act as a magnet for migrants.

The economy suffers from a lack of employment, so a large informal sector has developed (e.g. garbage pickers): automation has compounded the unemployment problem. However, Mexico City accounts for 35% of GDP and it is the financial centre of the country. The economy is served by an inadequate transport system, the overland rail system is poor and 2.5 million cars clog the roads creating a 4 km/hour crawl. This combined with poorly regulated industry, creates a problem of pollution. The city exceeds World Health Organisation guidelines for many pollutants such as lead and nitrogen oxides. Mexico City generates 9 000 tonnes of domestic rubbish per day and there are worries over seepage contaminating the city's water supply. (5)

(ii) Solutions are hampered by the change of government ministers every 6 years: no long term policy exists. However, some solutions are being attempted. The problem of housing has been tackled by giving de jure ownership (legal rights) and hence security to once illegal settlers. Self help and upgrading have been emphasised, e.g. San Miguel residents have dug drains with the help of government money. Transport policy is seeking not only to help movement but to help decrease pollution: the Metro (built since 1967) now makes over two thousand million passenger journeys per annum, and 'green' buses are used; there are now controls on certain car licence plates, on certain days.

Answers to Unit 8

Question	Answer	Mark

The main government policy, begun in 1981, is one of decentralisation, encouraging development away from the Federal District. There is a buffer zone or green belt to the south of the city and new industry (the main cause of migration) is prohibited within the Federal District. However, it is still to be seen if this will be a success: lack of money and other resources, and the huge national debt, make improvement uncertain. (5)

Examiner's tip The candidate has made it clear that lack of money is the root of all the problems and other difficulties are just symptoms of this. As it is such a key point, it is appropriate that this has been left to the end of the answer.

---

2 (a) (i) Medical services have been rationalised, reducing the service and the number of locations in which they are found. (2)

(ii) The shops have been concentrated into selected settlements although the number remains the same. (2)

Examiner's tip Two marks are available for each so two points need to be made.

(b) First reason: To save money. (2)

Second reason: People are becoming increasingly mobile (private car ownership), therefore you do not need services everywhere. (2)

(c) First reason: It is on the main bus route so it is accessible to people. (2)

Second reason: It already has many services, so it is easier and cheaper to expand, rather than moving existing services. (2)

Examiner's tip It is important that the two reasons are as different as possible, otherwise full marks may not be allocated.

(d) First reason: The inhabitants do not currently have to travel outside the village for most of their services. (2)

Second reason: People without cars, e.g. young and low income, will not be able to access services after the changes are made (no bus service). (2)

(e) (i) Central Place Theory (CPT), through its concepts of catchment area and hierarchy, would help suggest the third village (probably Boxford). (2)

(ii) Shops need to meet their thresholds, so if there are not enough people within range (the distance people are prepared to travel), shops cannot remain profitable. (2)

Answers to Unit 8

Question	Examiner's tip

3 (a) (i) A definition is all that is required. In this case, the process of the wealthy buying up properties and upgrading them.

(ii) Several wards could be selected, they have to be near the CBD (access to jobs) and have a low percentage of overcrowding and a low percentage lacking a WC. Number 6 would fit this description.

(iii) Gentrification will occur when the necessary conditions exist. These are: large (usually Victorian) premises that are structurally sound and economic possibilities to invest and make profits: in the UK the availability of improvement grants in designated general improvement areas brought increased gentrification in the 1980s. Central city locations are becoming more desirable because of increasing commuter congestion which can be avoided (stress, time and money).

(b) (i) Fig. 4 shows that households with poor WC provision tend to be in central or eastern (and NE) locations. Overcrowding appears to be in the central city and the north east towards Leith Docks as well as in the far east such as Liberton. That is, there is much overlap between the patterns.

(ii) This is the location of older housing, and therefore fewer facilities. It was built for the 'working class' in Victorian Scotland. Additionally, these 'inner city' areas house the lower income residents who cannot afford to upgrade their homes.

(iii) For area 23, this is possibly an off-centre, post-war council estate where small housing units (probably flats) create high densities. District 19 is an old part of the city, with high density Victorian housing (tenement or back-to-back).

(c) You need at least one detailed example (e.g. Cambridge and its hinterland) and you must explain what suburbanisation of the countryside is. You must consider the extent of the positive social and environmental consequences that are seen, but 'extent' infers that there maybe negative aspects also. Issues to be discussed include: new jobs and housing in the villages, commuter patterns and their problems, the decline of traditional jobs (farming) and services and the effect on village community spirit, amongst others.

4
- You need to explain and define multiple deprivation (remembering that it is a relative term), and what an inner city is (not necessarily found in the central city).

- There are two vicious circles that need to be broken if multiple deprivation is to be stopped. The first relates to government finance (lack of industry - low tax base - low expenditure on services - unattractive environment - lack of industry). The second is concerned with the individual (poverty - poor housing - stress and strain - poor education - lack of skill - no job - poverty). This cycle of poverty is self-perpetuating and continues from one generation to the next.

Answers to Unit 8

Question	Examiner's tip

- The next question you must ask is 'Why is it difficult to lower poverty concentration?' You must bear in mind that those who can succeed will leave and those areas that need help most cannot afford it and so 'rumps' are left.

- Finally, you must cover government attempts to counter this problem (using good examples such as LDDC) and discuss whether or not they are successful.

5 (a) (i) and (ii) These are skill based questions. Both require the use of figures and dates but not necessarily reasons.

(b) The bulk of the marks here are for reasons explaining the links between urbanisation and economic development. Answers need to refer to the shift from primary (especially agriculture) to other industrial sectors and how this is linked to wealth.

(c) (i) and (ii) These questions are primarily concerned with counterurbanisation. However, candidates must remember to address cities both growing and shrinking in population and make reference to both periods, in order to show the slowing of counterurbanisation. It is necessary to use examples, as the question says '*and using your own knowledge*'.

9 POPULATION AND RESOURCES

Question	Answer	Mark

1 (a) (i) There was an ever-decreasing population in Inner London between 1961 and 1976, reaching a low of -20. This figure then changed slightly to -16 by 1976 to 1981, but there was then a dramatic change to a loss of only -2 by 1981 to 1986 that is, the loss of population appeared to have stopped or at least slowed. (3)

Examiner's tip Good use of figures and dates here is essential. Description of the change over the period is all that is required.

(ii) This was due to the process of counterurbanisation characterised by increasing diseconomies of scale associated with large cities, including: the rising costs of land prices (for firms and homes), increased congestion and pollution costs. There was a loss of traditional jobs (e.g. London docks) and jobs relocated to green field sites so people followed. Also better transport allows greater mobility and now this is possible people prefer to live in a pleasant rural setting. (4)

Examiner's tip At least four reasons are needed (hence the mark allocation). The candidate has made quite a good use of examples and has included the necessary terminology.

Answers to Unit 9

Question	Answer	Mark

(b) (i)

[Graph showing Population shift from -20 to +20 on y-axis, with time periods 61/66, 66/71, 71/76, 76/81, 81/86 on x-axis. Remoter rural districts (●) points: approximately +2, +6, +11, +7, +7. Principal cities (○) points: approximately -14, -12, -12, -10, -6. Note: 76/81 and 81/86 remoter rural points shown as ×.]

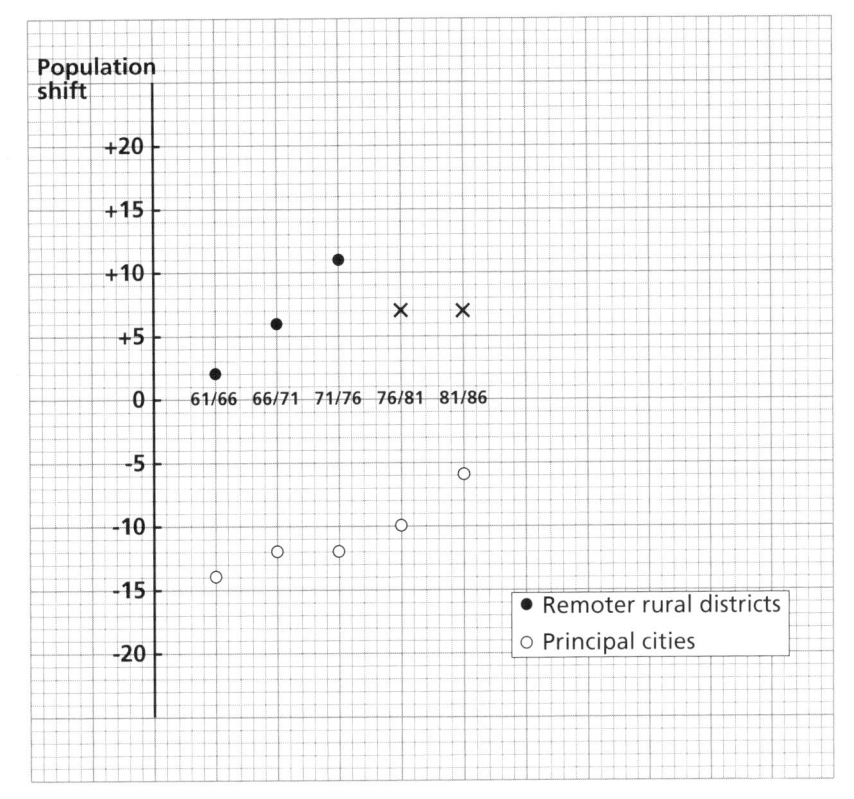

(2)

Examiner's tip This is a skill based question so make sure that you use the equation given and plot precisely on the graph.

(ii) 1. Compared with the national trend, large cities had falling populations because of the disadvantages (e.g. congestion) of living in them. This peaked in 1971–76, a period when job shedding occurred due to rationalisation in older (19th Century) industrial areas generally found in the centre. Also the perception of the inner city deteriorated (e.g. St. Paul's riots in Bristol). Towards 1981–86 this fall steadied, as the population neared an equilibrium optimum population and government policy was directed to help these areas (e.g. the Lower Don Valley redevelopment scheme in Sheffield). (4)

2. People moved to remoter rural districts because of the availability of personal transport and the better quality of the environment and standard of living. People also followed employment opportunities, as firms moved out of traditional manufacturing areas because of spiralling costs, to those areas with cheaper costs. (4)

Examiner's tip Remember the question is asking you to give possible explanations, not simply describe the patterns.

(c) (i) Up to and including 1971–76, industrial areas enlarged their populations considerably above the national trend. This was because of industrial expansion in the UK. After 1976, this trend slowed primarily because of the global recession that affected UK employment. Also, as infrastructure improved and the nature of industry changed (to 'hi-tech' and 'footloose'), firms had a greater freedom to move elsewhere, thus reducing the population of industrial areas still

Answers to Unit 9

Question	Answer	Mark

further. By 1986, industrial regions were actually losing people in relation to the national average, because their populations remained stationary. (4)

(ii) New towns had a rapidly increasing population in the early years (1961–71) as a result of government policy (e.g. relocation from inner city clearance schemes). This increase slowed, but never fell below the national average increase in population, because of the natural advantages new towns had (e.g. cheaper land values) combined with government incentives. These incentives such as Telford's Enterprise Zone, are still available. These advantages create jobs, so people move to them. Finally, migrants are generally younger, so new towns generally have a higher fertility rate. (4)

Examiner's tip Refer back to the table. Again you must give explanations for the changes including examples.

2 (a) (i) A = crude death rate (CDR) ; B = crude birth rate (CBR). (2)

(ii) Rate per 1000 people. (1)

(iii) Natural population change (in this case natural increase). (1)

(iv) Stage 2 = early expanding; Stage 3 = late expanding. (2)

(b) Graph 3, because both 1 and 2 show the CBR is lower than the CDR. Therefore the population would be decreasing. This is not the case for most developing countries which have rising populations as the CDR is falling (due to eg. better health care). (3)

(c) Developing countries usually fall into the second and third (early and late expanding) stages of the demographic transition model, whereas developed countries typify those countries that have entered stage 4 (low fluctuating), or have even reached a point where crude death rates have exceeded birth rates. For example, Ethiopia in 1989 had a CBR of 48 and a CDR of 18 giving a natural increase of 3.0%. The UK had a natural increase of only 0.1% in 1989, and Sweden had a negative rate of population growth as its CBR was 12, whilst the CDR was 13, giving a decrease of 0.1%.

Examiner's tip A detailed answer is expected because of the mark allocation. Several reasons for the differences in natural population increase are needed and given. The use of figures is essential, using the term 'high' will not achieve as many marks.

As economic development proceeds there is a decline in mortality as the countries pass through the epidemiological transition (changes over time of the causes of death), and fewer people die of endemic (e.g. yellow fever), parasitic and dietary deficiency diseases. This is because higher standards of living result in more being spent on food, clean water and medicine. For a period the CBR remains much higher than CDR because children continue to have an economic value as a cheap source of labour and because there is a delay in people's perception in the decline in the CDR. They continue to have many children to offset their perceived view of a high rate of infant mortality. In developed countries, later marriage, higher levels of female employment and higher status, availability of contraception and the economic burden of children, have all led to a sharp fall in the CBR. (9)

Answers to Unit 9

| Question | Answer | Mark |

> **Examiner's tip** A thorough list of possible reasons for changes has been given. In particular, the candidate shows good understanding of the time lag between people's perception and population reality.

A final reason for the higher growth of developing countries is their population structure. The developing countries in 1985 had 37% of their population under 15, whereas the comparable figure for developed countries was 22%. Thus, even if fertility drops to two children per couple the population will continue to grow for many years due to population momentum.

(d) Regions of sparse population density generally have physical environments which are hostile to people and their economic activities. Examples of such regions are deserts, mountainous areas, polar regions and areas of dense tropical forests. In these locations opportunities for intensive agricultural productivity are low and poor accessibility hinders resource exploitation and manufacturing, hence the carrying capacity of the land is low and the population densities will reflect this.

There may be a localised resource that offsets the overall constraints of the area. Egypt, for example, is largely desert, but bordering the Nile very high population densities (over 50 per sq km) are found, in marked contrast to the surrounding desert where under 1 per sq km is common. This is also true of oases where intensive agriculture can be practised. Similar pockets of dense population also exist where valuable mineral resources are discovered. Provided the resource is large and of high value it is worth establishing a town. The link between the land's ability to provide food and the needs of the population is broken as food is now brought in. Good examples of this are: the development of Kiruna in Sweden where vast iron ore reserves exist, the exploitation of gold reserves in the inaccessible forested mountains of Papua New Guinea, and mining settlements in Siberia. (7)

> **Examiner's tip** Identify the reasons why some areas have sparse populations, and then use a case study to explain why pockets of higher populations exist. Without the examples, your answer will receive low marks.

3

> **Examiner's tip** Look at the wording carefully! This question deals both with the limitation, and more unusually, the growth of populations. If you fail to recognise the need for some populations to grow, you are unlikely to achieve better than a grade C. You have to describe both the policies adopted and give reasons for their adoption. You will need to identify a set of contrasting countries and governments to exemplify your points.

Governments may wish to manipulate the growth and limitation of their populations for a range of economic and social reasons. They can achieve this by direct or indirect policies. Populations change by natural increase and by migration; both may be targeted by government policy. Most frequently discussed are the difficulties encountered by developing countries which are passing through stages two and three of the demographic transition and may wish to accelerate the decline in the CBR. Another group of countries actually

Answers to Unit 9

Question	Answer	Mark

wishes to enlarge their population size. These typically fall into three categories: countries in the 'New World' which are underpopulated and are unable to exploit their resources, those where for strategic or ethnic reasons a government wants to see an increase in population, and those where an aging population, and therefore a reduction in the labour force, is seen as a problem.

> **Examiner's tip** This type of introduction identifies the key points, serves as a framework for the development of your essay and, if you were to run out of time, shows the examiner you were set on the right course.

Developing countries often have difficulties with population growth that is running out of control. The types of difficulties a government may be facing are lack of resources, in particular food, which leads to malnutrition, famine and probably political instability. A reduction in population expansion may improve living standards, though economists such as Boserup would not agree with this premise.

China exemplifies a country that has taken a series of direct measures to reduce the birth rate. It has 22% of the world's population, with most of its population concentrated in relatively small areas of fertile lowland. The death rate is extremely low, 6 : 1000, and life expectancy is 70. The birth rate for a developing country is relatively low at 21 : 1000 but despite this, the population is expected to grow by 100 million over the next 6 years. The Chinese government has operated a family planning campaign for the last 25 years (the much publicised one child policy) and has succeeded in reducing the birth rate dramatically. The average number of children per family is now 2.4, and is lower in urban areas 1.7 (in 1987). China has achieved this by a well organised system of financial rewards and penalties, combined with a strong degree of enforcement of late marriage and strict registration of children: there is undoubtedly extreme pressure on women to have any second pregnancy terminated.

Not all governments are able to enforce policy as effectively as China, but a number of Asian countries have reduced birth rates by direct incentives and disincentives: Singapore gave priority in schooling to those with two children and combined this with tax incentives; in Thailand family planning was provided free; and in Bangladesh participants in family planning schemes were given a free sari.

> **Examiner's tip** The candidate has supplied some excellent examples and accurate figures.

There is some evidence to suggest that indirect methods of reducing fertility may actually be more effective. Fertility correlates strongly with education levels, employment prospects and the status of women. Countries that have 'modernised' have often found a drop in fertility as women become part of the workforce. The Demographic Transition Model suggests that economic advancement is the driving force behind the fall in CBR. Consequently, the very poorest countries, such as Nepal, Senegal and Bangladesh, even after adopting family planning policies, have not always seen a fall in CBR, so some degree of socio-economic development may be essential.

> **Examiner's tip** Here the candidate raises the easily-overlooked yet vital link between economic wealth and family size, showing true depth of understanding.

Answers to Unit 9

Question	Answer	Mark

The countries that have adopted pro-natalist policies tend to be developed, and are motivated by concerns over geopolitics, ethnic groups, and labour force. A good example of a government adopting pro-natalist policies is France, where the BR has been low for many years, they offered some cash incentives to have more children and if a woman had six or more children she was given a medal! Romania under Ceausescu also adopted stringent measures to try to increase the CBR. Abortion was restricted in 1974; financial pressures were imposed on unmarried people; and charges could be brought against women who were pregnant and didn't produce a baby after 9 months. Another former communist country trying to boost population by a series of incentives was Hungary. However, these techniques were not very successful as they ran counter to individual family economics. More effective were the methods adopted by countries like Australia where immigration was encouraged in the 1950s and 1960s to try to increase the population, to allow them to develop their resources. (23)

Examiner's tip As has been mentioned, you must include reasons and examples of why some countries actually want to increase their populations.

---

4 (a) (i) 1. This is because population increases factorially; if a couple have four children it is assumed that each, with their partner will also have four children. Therefore, the population in this case would double every generation (1,2,4,8,16). (2)

2. Food supply only increases due to an increase in inputs (e.g. labour) and these are only added when available. A previous use of inputs has no bearing on later outputs, unlike preceding generations of people. (2)

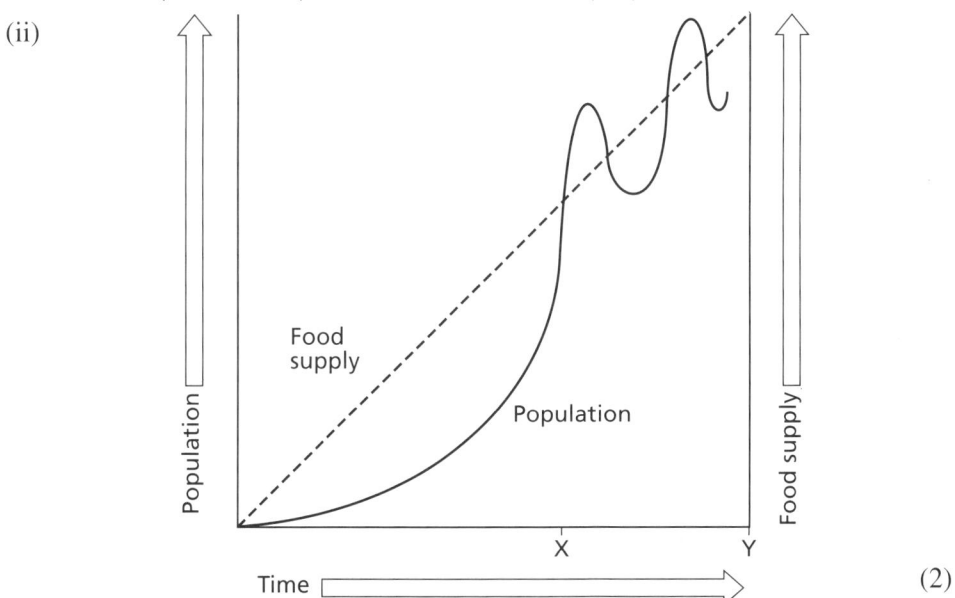

(ii)

[Graph showing Population (solid curve rising steeply then oscillating) and Food supply (dashed straight line) against Time, with points X and Y marked on the time axis] (2)

(iii) Population will exceed food supply (in part due to a time lag in births), so malnutrition will occur in the population and it will become more susceptible to disease. This persistent famine eventually leads to starvation and epidemics, increasing the death rate; the population 'crashes'. Now that the population has fallen below that which can be sustained by the food supply, it will begin to rise once more, repeating the cycle. (3)

Answers to Unit 9

Question	Answer	Mark

(iv) The model only takes into account present, discovered, or accessible resources. Technology is assumed to be constant. Both of these ignore future discoveries that can change the resource-population ratio. Malthus did not account for different social, economic or cultural aspects of places (e.g. Muslim societies) and he did not anticipate that population growth might not continue to show a constant geometric increase, as populations reduce their birth rates due to improved economic well-being. (4)

(b) (i) A country can increase food imports or cut back on agricultural exports, diverting their supply to domestic use. Food aid is the most immediate help. (2)

(ii) The country needs to restructure its farming, producing a new agricultural system. This might include intensification (e.g. use of chemicals and machinery), expanding the land area farmed and using improved technology such as genetic engineering of seeds and animals (i.e. the Green Revolution). (3)

(c) (i) If there is under-population, the resources are not fully utilised compared to their most efficient optimum (e.g. parts of Australia). If the birth rate drops and there is an aging population, the economy may suffer, as the working population decreases proportionally and has to support a greater dependent population (e.g. Italy). Finally, there may be political or strategic reasons such as isolation from the global community (e.g. Libya). (3)

(ii) If there is over population (too many people for too few resources) this causes 'diminishing returns' from the given resources, leading to shortages. This is commonly associated with 'running to stand still': a country frantically developing its economy simply to feed and house the ever growing number of people. A country may have to discourage population growth because of conditions placed upon it by supra-national agencies (e.g. World Bank). (3)

10 INDUSTRY

Question	Answer	Mark

1

Examiner's tip This question appears at first to be quite confusing. Do not panic! It is vital to take time to analyse the key words. The first part refers to industrial inertia, whilst the second is about agglomeration, linkages and external economies of scale.

Industry is not an abstract phenomenon, but occurs at places; one must ask, therefore, why does it locate where it does? Its location may be for a variety of different reasons, including chance (e.g. Morris [the car manufacturer] lived in Oxford), but is more often to do with economic influences.

Examiner's tip Asking a question is a good way of opening an essay if the candidate is confident of handling the material.

Industrial inertia is where fixed capital investment (factory buildings) encourages industry to maintain its original location, even though the initial reasons may have disappeared. For example, Sheffield is still world famous for its high quality steel products (the Stocksbridge Steel Works produce high

Answers to Unit 10

grade metal for the aerospace industry), even though the original raw materials of coal and iron ore have long since been exhausted in the area. There are a number of reasons for this inertia. First many of the companies in the Sheffield area, such as Sheffield Forge Masters, have invested many millions of pounds in their buildings and machinery. This investment is immobile, and if a new location, such as a coastal site with easy access to foreign raw materials was chosen, much of this investment would have to be 'written off'. Also, over the past 150 years, Sheffield has acquired a pool of experienced, skilled workers (apprenticeships etc). The cost of either transplanting this (if the workers were prepared to travel) or educating and training a new labour pool would be prohibitive.

> **Examiner's tip** The use of Sheffield serves two purposes. Firstly, you are required by the question to use detailed examples, of which this is one! Secondly, by putting the abstract idea of inertia into a real place, it illustrates understanding.

Industrial inertia also includes prestige: an area being well known for a particular product. In the case of Sheffield, high quality steel products, and particularly cutlery, are its forte. Companies such as Richardson's proudly display the 'Made in Sheffield' logo on their products; a deliberate marketing strategy, as consumers know the quality of the area's products.

Once industry has concentrated in a particular area, there are strong reasons why new industrial investment is attracted to it. If one looks at Weber's industrial location theory, he states that firms will take advantage of skilled labour pools (discussed above) if the savings made by this labour force outweigh any increased transport costs to reach it. Below is an illustration of agglomeration or external economies of scale. That is, taking advantage of factors that occur outside the individual firm.

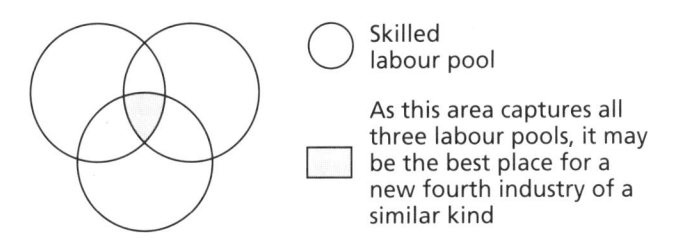

○ Skilled labour pool

▫ As this area captures all three labour pools, it may be the best place for a new fourth industry of a similar kind

> **Examiner's tip** The use of sketch diagrams will always help explain complex ideas. It is important to make a written comment referring to the diagram rather than simply inserting it for aesthetic value!

Industrial linkages also forge industrial concentrations. These, in simple terms, are where associated activities benefit from being close to other, similar, industries. For example, the Toyota Corporation centres nearly all of its production in Japan in and around Toyota City in the Nagoya region. This is not only in a central position on the Pacific Coast with excellent transport links, but it also has a large and reliable workforce. The main assembly lines are supplied by thousands of sub-contractors supplying different parts (e.g. tyres, gearboxes etc). This enables 'just in time' practices to be used, thus saving on storage costs.

Answers to Unit 10

Question	Answer	Mark

The petrochemical industry is another good example of industrial linkage. The materials for petroleum associated industries are supplied by oil refineries through backward linkages, whilst forward linkages exist between petrochemicals and the processes it supplies such as plastics. Such an example can be found at Llandarcy, South Wales.

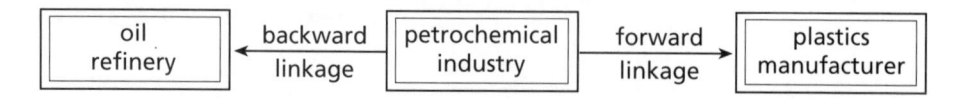

```
┌──────────┐  backward   ┌──────────────┐  forward   ┌──────────────┐
│   oil    │ ←────────── │petrochemical │ ─────────→ │   plastics   │
│ refinery │   linkage   │   industry   │   linkage  │ manufacturer │
└──────────┘             └──────────────┘            └──────────────┘
```

Industrial linkages and external economies of scale can lead to energy savings, research and development cooperation and a strong political bargaining position for the area, as well as reduced transport costs. Myrdal's model of cumulative causation demonstrates industrial concentrations due to the presence of a successful industry in an area. Its success creates demand for service industries associated with it; a case of 'success breeds success'. Many urban areas are a direct result of this.

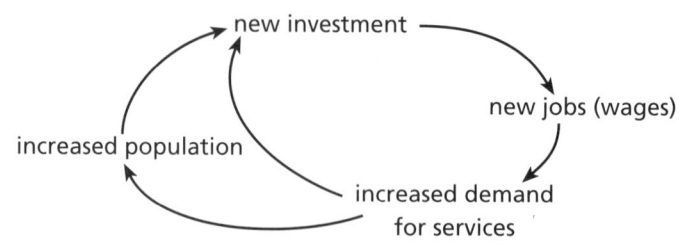

```
              → new investment ─
             ↗                    ↘
                                    new jobs (wages)
                                        ↓
    increased population
         ↑                       increased demand
          ←──────────────────    for services
```

Of course, there are other locational factors such as government policy, or managerial desires. Is it a coincidence that Toyota's Derby plant is near a golf course?

The future is uncertain. Will the present location still pull new industry to it and thus maintain the location through inertia? As (tele)communications continue to improve we may see, especially in the Developed World, a further dispersal of industry to new locations away from present industrial areas i.e. counter-urbanisation and 'footloose' industry. It is possible to envisage a world where present and past industrial locations become less important to newer types of industry. (24)

> **Examiner's tip** Whenever you are writing an essay, you must be conscious of style and include an introduction and conclusion. In this answer, the candidate has concluded by exploring possible future occurrences. This is not stated in the question, however, some evaluation is always helpful.

Question	Examiner's tip

2 (a) This first point requires you only to explain the key ideas of Weber's LCL model. You must cover the following: the importance of transport costs including the MI and the Varignon Frame, labour costs (in particular the relationship between saving labour costs compared with the increase in transport costs, i.e. the critical isodapane) and agglomeration economies. Use diagrams to illustrate your points

(b) This asks you to evaluate the theory's relevance today as it dates from 1909, as well as challenging some of the basic assumptions (e.g. proportional travel

Answers to Unit 10

Question	Examiner's tip

costs over distance). This depends very much on the nature of the industry being studied. Heavier industry such as steel still locates where raw material can be accessed most cheaply, but newer, hi-tech industry in the quaternary sector is relatively 'footloose' and is able to locate in most places (although they may take advantage of certain factors such as universities). Examples of case studies need to be used to illustrate both Weber's applicability and irrelevance. The question enables you to sum up the evidence for and against Weber, in particular in your conclusion.

3 (a) (i) and (ii) These are simple skill based questions. Remember to read off from the correct axes!

(b) (i) You must give reasons as to why this has decreased. A one liner will not score four marks.

(ii) Again, thorough reasons need to be given to explain this country's continuing de-industrialisation.

(c) (i) and (ii) Description (i), and explanation (ii), have been separated in this question and you must also do so in your answer. Part (i) refers to tertiary sector expansion, whilst (ii) is based on the technological revolution (telecommunications) and the increasing wealth of the country.

(d) (i) This needs a thorough definition as it is worth three marks. Give an example.

(ii) This is linked to increased wealth and particularly to the expansion of telecommunications.

4 (a) (i), (ii) and (iii) All these are skill based questions, read the information on the graph and the instructions carefully.

(b) You are required to give a thorough definition and distinction between the two with reference to an actual firm.

(c) This question is about how large multi(trans)national companies have expanded globally, locating near to new markets, thus reducing transport of bulky vehicles. They also take advantage of local incentives and cheaper workforces around the globe. Specific examples are necessary, such as Toyota's expansion in the UK.

5 (a) (i) You must, of course, understand the Location Quotient. That is, if the figure is below 1 it is less than the average and above 1 it shows a concentration above the average. The question wants you to explain how the LQ helps us to understand the distribution of the woollen textile industry, including agglomeration economies.

(ii) This question evaluates the role of the Government i.e. interventionist policy against free-market ideas. Issues such as lame ducks - infant industry need to be discussed, as well as broader social issues rather than only economic ones. Finally, the question asks you for an opinion, make sure you have weighed up both arguments first.

Answers to Unit 10

Question	Examiner's tip
(b)	You need to base your answer around one manufacturing industry only e.g. the motor industry and have a good knowledge of what has happened to this with reference to its location, organisation and size in recent years. You need to know how electronic communications have allowed the centralisation of information and the decentralisation of operations. The role of TNCs needs to be examined and their dominance over national firms, and even countries. As hinted, new revolutions in the money markets (e.g. the Big Bang) has increased the international mobility of capital, thus enabling the TNCs to shift their investment locations.

11 AGRICULTURE AND FOOD

Question	Answer	Mark
1 (a)	Often a subsistence form of agriculture, where people farm an area for a limited period before moving on to another plot (i.e. not permanent) e.g. the Fulani in W. Africa.	(3)

Examiner's tip For 3 marks the definition must be thorough, with an example.

(b)	a = underutilisation; b = optimal; c = overutilisation.	(2)
(c)	- a reduction in the fallow period; - a long period of drought.	(2)
(d)	In general terms the system needs to become more intensive. <u>Irrigation</u> would enable all year farming and hence more biomass would be produced. <u>HYV or new breeds</u> could be introduced which yield more output per hectare. Most important would be to add <u>fertilisers</u> (domestic, animal or artificial) so that the land's fertility would be maintained even though more would be demanded from it.	(6)

Examiner's tip By underlining key words, the candidate is stating his/her 'three ways' as the question asks. It is necessary to explain how each can improve the food supply without soil fertility reduction as there are 2 marks allocated for each method.

(e)	This would depend on the population size in relation to the land available. Further south in W. Africa, the fallow period would be reduced.	(2)

Question	Examiner's tip
2 (a)	In this first part of the question you simply need to be able to describe, with good examples, the main features of plantation agriculture (e.g. large organisation, often owned by foreign companies etc).
(b)	This requires you to give equal weight to both positive and negative consequences of plantation agriculture. Debates will centre over environmental concerns (i.e. monoculture in a tropical environment), social (land ownership etc) and economic issues as well as production yields. Do not look only at the problems, as many candidates may be tempted to do, as this limits the scope of your answer.

Answers to Unit 11

Question	Examiner's tip
3	This question is centred on how relevant von Thünen's model of rural land use is, which emphasises farming through economically rational decisions. It gives you scope to use other influencing factors on a farmer from all areas of the world, but as the question insists, you must use examples to illustrate your arguments. The government's role could be explored under economic reasons, although one might argue that these themselves may not be economically rational (CAP)!
4 (a)	This question has two parts to it. The first requires you to outline SE Asia's monsoon climate and the reasons for it. However, weaker candidates might struggle to link the area's climate to farming practice and how the peasant farmers take advantage of it. Obviously good detailed knowledge of both the climate and peasant farming are necessary.
(b) (i)	This is quite straightforward, all you have to do is outline the main developments of the GR (HYV etc) and refer to specific examples of these, both in research and implementation.
(ii)	This part requires an evaluation of the success of the GR and, as with any evaluation, all aspects should be covered i.e. has the GR been successful or not, and if so, for whom and why? Are there any groups that have been adversely affected? Again the question asks you to use specific examples.
5 (a)	This is a resource based question in which you, after reading the advert, should highlight the cattlemen's views on the environment. The word 'consider' allows you to challenge the cattlemen's advert as possible 'propaganda', but, as with all evaluation, both sides of the argument should be presented.
(b) (i)	This may appear intimidating to a 'non-artist' but the skill of annotated diagrams does not lie in drawing, but rather on labelling technique, emphasising (in this case) the most important aspects of the agricultural system you have chosen as your example. Remember to make your labels clear and concise, using arrows to point to the features, enabling your words to occupy the margins of the page.
(ii)	This asks you to describe and explain how the environment interacts with the farming system that you chose in part (i). Do not consider a new system! This illustrates why you should read the whole question before starting any part of it. A good answer must consider all parts of the question, and give a description and explanation of both constraining and promoting aspects.
6	• This question demands that you evaluate the role of the natural environment and human factors in explaining famine. The argument centres around whether famine is caused by an uneven distribution of resources, rather than a lack of resources per se, which may be caused by climatic variations (primarily drought), as seen in Sahelian Africa.
	• Other issues that need to be discussed are whether local people are to blame (over population, overgrazing and thus land degradation and desertification) rather than external people (such as colonial and neo-colonial relationships, taxes, international trade etc).

Answers to Unit 11

Question	Examiner's tip

- All of these points need to be illustrated by detailed examples and put into the context of places.

12 TOURISM AND LEISURE

Question	Answer	Mark

1 (a) The expansions in the La Filette town area will be:

(i) new tourist development

- a new hotel and chalet complex is to be built on reclaimed land over Set Net Reef
- Marie Islands will be connected to the mainland by a causeway, thus enabling holiday villas to be built
- new apartments to be built on the town's vegetable plots

(ii) new port facilities, which include

- a dock at Fisherman's Beach
- a marina created by an artificial harbour
- dredging of the sand banks and sea grass areas

(iii) new industrial area near the proposed docks, including a sugar mill

(iv) upgrading the transport network, primarily linking La Filette by a motorway to the rest of the island

(v) expansion of the town itself, to accommodate the new workers etc.

Some environmental impact will be felt (although reduced to a minimum), such as a reduction in the extent of the coral reefs, changes in the sand/beach distribution, rescinding the nature reserve's status and river bank improvements. (6)

Examiner's tip The format this response takes is important as it is designed for press consumption. It should be easily digestible so point form is appropriate.

(b) Any evaluation of a development needs to include some kind of Cost-Benefit Analysis. However, this has its own inherent weakness, how can you quantify the unquantifiable? In this case what monetary value do you put on, for example, a coral reef or aesthetic beauty?

The potential environmental impacts are far-reaching and the initial question is how will the local ecosystem cope? An ecosystem is an assemblage of living and non-living elements which combine to create an integrated, functioning environment. The coral reef is such an ecosystem, which is very sensitive to pollution and interference, as the photosynthesising cells can only survive in 'clear' water. Coral is the basis for life in the system, acting as habitat (e.g. conga eels) and food (e.g. parrot fish), if it is removed, the dependent life will be lost too. Tourists themselves will add stress to these systems (coral 'picking' etc) and the economic developments will cause direct destruction (e.g. Set Net Reef) and create pollutants, such as sewage from the new hotels and industrial pollution from the sugar mill. If these stresses exceed the critical threshold of the reef system, not only the coral, but all the marine life that depends on it will disappear. This breakdown could also occur because the coral that is left may

Answers to Unit 12

Question	Answer	Mark

not be diverse enough to be viable (e.g. reproduction). If one part of the interdependent chain is broken, the other parts cease to function. This is compounded by feedback mechanisms.

> **Examiner's tip** Here the candidate has used a learnt definition (ecosystems) and applied it to the information given in the question (coral reef system). This connection will score high marks.

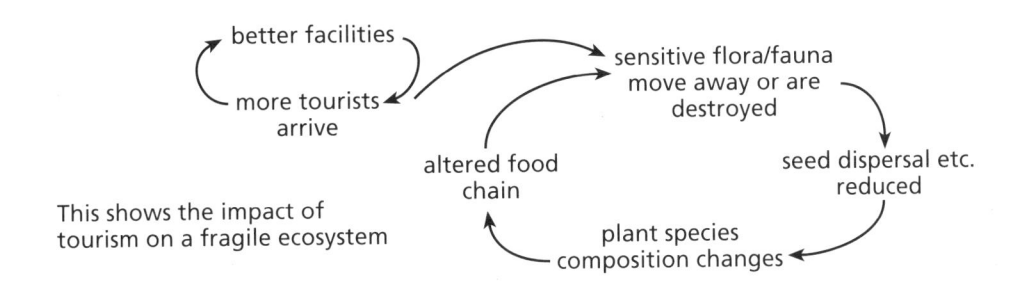

```
        better facilities
           ↗         ↘           → sensitive flora/fauna
                                    move away or are
           ↖ more tourists ↙        destroyed
              arrive                      ↓
                        altered food        seed dispersal etc.
This shows the impact of    chain              reduced
tourism on a fragile ecosystem   ↖  plant species    ↙
                                    composition changes
```

This shows the impact of tourism on a fragile ecosystem

The removal of the nature reserve will do nothing to help the situation, although one could argue whether development in a 700 metre stretch will make that much of an impact on the coral ecosystem. Other environmental concerns such as rubbish (e.g. non-biodegradable plastic bottles) left by the extra number of people or footpath erosion will arise.

Landform processes will be altered; in particular, long shore drift. The predominant direction of coastal sediment movement appears to be from northwest to south east. The new port and hotel/chalet complex will trap sediments on their western sides. Thus, beaches will be created on the western side of the causeway and artificial dock walls. This could be beneficial for sunbathers at these locations, but it means that less sediment (sand) will travel south eastwards. Therefore, the Golden Sands Hotel will be robbed of its golden beaches! If beaches are removed in places this will increase the waves' erosive force in these areas as there will be no beach to shallow the depth and absorb wave energy. The reduction in the coral, sand and seagrass areas will have a similar effect; at present, they act as a barrier protecting the shoreline. Plunging (destructive) waves will result, rather than spilling ones.

Straightening the river and reinforcing its banks will affect channel flow and it appears that less sediment will accumulate at its mouth. Finally, where the vegetable plots are, the slope will be cut back so that a flat area for the villas and chalets will be created. This may mean that the movement of water through and across the land will change. With a steeper incline where the high ground meets the flat, low ground, one would expect more surface run off and possibly more erosion. In turn, this may cause slumping of the slope.

> **Examiner's tip** As requested, the answer refers to landform processes. It is better to structure the essay by topic, rather than tour the diagram stating the impacts of each development.

In economic terms, the developments should bring many benefits, the most obvious of which is employment. If people have work, this means that the general standard of living will improve as there is more money for services etc. Job opportunities would not only include the initial construction, but also hotel staff

Answers to Unit 12

| Question | Answer | Mark |

etc after completion. The industry will, hopefully, reduce any overdependence on tourism that La Filette might develop. Value will be added to local primary products, thus increasing foreign currency earnings. More important will be the increase in tourists who will spend their currency in La Filette. Through increased taxes etc the government will be able to spend money buying foreign technology and invest in infrastructure (physical and social).

Not all of this money will stay in the country, as leakage will occur. The World Bank estimates that 50% of tourist spending in Developing Countries reaches the Developed World in some way. Tourists often demand that their lifestyles should be imported with them, so food and furnishings have to be bought from abroad. Much of the tourist development will be by foreign firms, who will own the hotels, airlines etc. Many of the workers, such as Cordon Bleu chefs and Tour representatives, will be foreign.

Tourism will increase the prices both of goods and land and in this case, locally, the removal of the vegetable plots may lead to an increase in food prices. This artificial inflation may force locals out of the area. The conspicuous lifestyles and consumption may cause resentment (e.g. crime), heightened by the loss of local access to facilities, such as the best beaches reserved for the tourists. Jobs associated with tourism are seasonal (high season is Nov - Feb in the W. Indies) in their nature and so job security is low, and many of the jobs will be low-paid e.g. cleaners, whilst other jobs are servile in their nature: white tourists being waited on by black workers, a throw back to colonialism.

> **Examiner's tip** In this question you are required to consider the 'impacts' of certain things. The better candidates will include both benefits and problems associated with the proposed developments.

The future is uncertain. Will La Filette become just another place in the 'pleasure periphery' of the Developed World, going out of fashion as quickly as it came in? Its development may in actual fact be detrimental, for example the natural beauty of the area may be spoilt or the fish may disappear, causing problems for the local fishermen.

It may become a magnet for the area, people may migrate in search of work but finding none, may resort to petty trading, prostitution or crime. The economic development may have spin off effects as more wealth and investment is attracted and so a more diverse base of economic activity may evolve. This wealth may then spread and diffuse into the other regions of the island, thus benefiting the entire country. However, the tourists may cause a 'demonstration effect' creating new behaviour patterns in the local economy e.g. the desire for American Levi jeans and Coke. However, the wish for employment and income and the potential to improve people's standard of living will probably outweigh the other considerations.

(17)

> **Examiner's tip** Weaker candidates might omit these final points. What are the possible economic climates that the La Filette proposal may bring?

Answers to Unit 13

Question	Examiner's tip
2 (a)	This question allows you to use either one detailed case study or several examples of a tourist development and its impact. It is important that all the areas (local population, communities and cultures) are considered. The tourist impact on the environment is not required, and if precious time is wasted here, it is lost from elsewhere!
(b)	Although the question does not ask for examples they must be given to help illustrate your answer. The main discussion here is that tourism is cyclical and fashionable. In the early stages of tourism, a place is seen as desirable by tourists, but as the life cycle of the product (tourism) matures to mass consumerism, the area becomes overdeveloped and spoilt, making it relatively unattractive, compared with new, virgin, exciting and now fashionable other places. This explains why, globally, tourism is expanding as new places open up to it, but, for a particular locality that goes through the cycle, boom and bust may occur. Other issues need to be discussed, such as the demonstration effect and the leakage of money from local economies, which may also explain why tourism is seen as a weak foundation for economic prosperity.
3	• This historical geography based question necessitates you linking the growth of British seaside resorts such as Brighton, to the expansion of the railways, increased wealth and more free time.
	• After considering the reasons for growth and using examples, you must also explain how actual places grew, and the facilities such as promenades and piers they provided.
	• In the second part of the question you need to use specific examples to explain what problems arose from this expansion, looking at economic issues such as transport, social issues such as class interaction and environmental ones such as over development of the coastline.
4	All parts of this question are straightforward, requiring you to use specific examples, although '*discuss*' implies not simply description but explanation and evaluation too. The better the knowledge displayed and the greater the depth of the case studies, the better grade your answer will receive.

13 SPATIAL INEQUALITY, REGIONAL DISPARITIES AND DEVELOPMENT

Question	Answer	Mark
1 (a)	This shows that some people believe that LDCs are a resource to be exploited for 'our' (DCs) benefit, whether it be as a pollution dumping ground or tourist resort. LDCs are very much 'Third World' i.e. third in order of priority, and as such have to play under 'our' rules of the capitalist monetary system. Hence, their resources are harnessed for 'our' monetary gain and 'they' (the LDCs) should be grateful for any, even the worst/most damaging industrial employment 'benefits'.	(5)

Answers to Unit 13

Question	Answer	Mark

Examiner's tip Although the major points are outlined, perhaps the candidate could have underlined or high-lighted them:
(i) That the world is for the <u>DCs benefit</u>
(ii) LDCs play by <u>our rules i.e. money</u>
(iii) LDCs should be <u>grateful</u> for any benefits

(b) (i) Gunner Myrdal (1957) used backwash to describe the flows of labour, capital and goods from the periphery to the core. The core grows at the expense of the exploited periphery as the terms of trade are loaded in its favour. (3)

(ii) This is the (un)intended consequence of an action related to either opening/expansion or closing/reduction of industry. If a firm expands this will create new employment, not only directly, but also in the firms that have linkages with it e.g. suppliers.

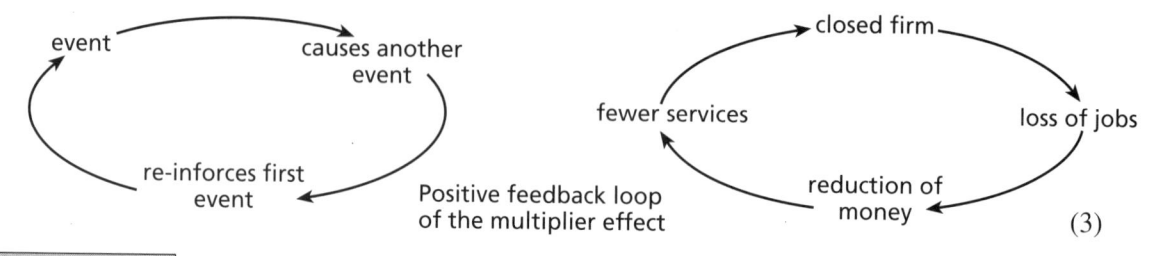

[Diagram: Left loop — event → causes another event → re-inforces first event (Positive feedback loop of the multiplier effect). Right loop — closed firm → loss of jobs → reduction of money → fewer services → closed firm.] (3)

Examiner's tip Two comprehensive definitions have been given. The flow diagrams show that the candidate truly understands the multiplier effect

(c) This describes how powerful states in the Developed World are able to retain or gain influence over underdeveloped countries, often former colonies. The countries dominated are not directly ruled but are apparently independent. Domination is for political influence or gain, and it is achieved through economic policy (trade and investment), for example, tied aid and Foreign Direct Investment. (4)

(d) **National**

Development does not occur everywhere at the same time or to the same extent, as Perroux, amongst others has commented. This is true at all scales, including all countries, the UK being no exception. In simple terms there is a North (periphery) - South (core) divide. However, regional disparities are seen as a problem for the government (a political vote loser if nothing else!) and so policy is used to help the poorer areas. Traditionally this can be achieved through 'carrots' (incentives) and 'sticks' (disincentives). Incentives include one off grants, loans, tax incentives and the building of factories etc in disadvantaged areas, whilst sticks are controls on industry limiting their activity in the core area. Carrots still exist in the UK, but sticks in the form of Industrial Development Certificates were abolished in 1981 (a reflection of Thatcher's free market ideas).

Examiner's tip 'Carrots and sticks' is a common analogy of government policy. It is good to see that the candidate has not fallen into the common trap of assuming that sticks still operate in the UK.

Answers to Unit 13

| Question | Answer | Mark |

In essence, there are three tiers of assistance. Supra-national aid is the first, mainly from the European Regional Development Fund (ERDF). Second, there is national help in the form of Regional Development Grants (RDGs) and Development Areas (DAs), and finally, local authorities give aid through bodies such as Enterprise Boards. However, since the 1980s there has been a 'rolling back' of regional aid in the UK, partly reflecting the recession, partly due to ideological thinking. There are three major schemes today.

RDGs are available in designated areas (either DAs or Intermediate Development Areas [IDAs]) which cover c. 25% of the UK population. Between 1960 and 1981 it is estimated that 800 000 jobs were created in these areas. DAs and IDAs also qualify for the ERDF which in the UK is spent mainly on infrastructure programmes. Development Agencies also work within these areas. The Welsh Development Agency has revitalised much of the Rhondda Valley, sometimes referred to as the Honda Valley, due to the success of attracting Japanese direct investment. Similarly, the Scottish Development Agency has been instrumental in Glasgow's Eastern Areas Renewal Project.

Also in Scotland, the Highlands and Islands Development Board has been helping this remote (from markets) rural area which is served by poor transport routes. Its main expenditure has been on building industrial premises e.g. for Flexible Technology on the Isle of Bute, as well as advertising and giving advice to firms. Through schemes it has created over 19 000 jobs (1977 - 1987). Since 1982 the Invergordon Enterprise Zone (EZ) has sought to attract employment e.g. the recording tape manufacturer Zonal.

Examiner's tip The use of these two detailed examples contrasts nicely with the general information that has been given. It is necessary to remember that the question asks *with reference to specific areas*.

EZs are normally located in inner-city areas that have experienced de-industrialisation and these are linked to Urban Development Corporations (UDCs). EZ incentives include rate exemptions, tax concessions and a 100% capital allowance for industrial buildings. UDCs have the powers to reclaim derelict land and to improve the environment and infrastructure. Loans and grants are also available.

The most successful UDC to date has been the London Docklands DC set up in response to the decline of the docks (loss of 18 000 jobs from 1966 to 1981). Although only £441 million of public investment occurred, this has attracted £4.4 billion of private money. 375 hectares of derelict land have been reclaimed and the transport system has been upgraded, including the Dockland Light Railway and the City Airport (plus the new M11 linkroad). Between 1981 and 1990, over 600 firms have created 15 500 jobs in the area.

Examiner's tip LDDC is a well documented example of government help to an area. This should not prohibit its use on the grounds that other people may be using it. Here the candidate has used the information to give a detailed account of what has happened in a specific area.

Teeside UDC, established in 1987, hopes to revitalise the Hartlepool - Middlesborough area (both have EZs). To combat high unemployment (c. 21% 1985) the Britannia EZ amongst other schemes including the Belasis Hall Technology Park, completed 6 500 m^2 of commercial, industrial and retail floor

Answers to Unit 13

Question	Answer	Mark

space, attracting 8 000 jobs. Other schemes have been used in the UK, such as Garden Festivals (Ebbw Vale 1992), and New Towns. Also, public direct investment has been diverted to peripheral areas e.g. Customs and Excise in Liverpool and Manchester.

Yet, despite all this help, the assisted areas are very similar in location today to those at the end of World War II. Therefore, one must conclude that the policy has not succeeded. However, without this help the situation in these disadvantaged areas would have been much worse. (18)

Examiner's tip Despite the fact that the question does not ask the candidate to evaluate policy, a good answer would include a brief judgement in the conclusion.

Question	Examiner's tip

2   This question is not difficult, but is does require you to be thorough. You must explore the rationale or basis for government intervention (including political, social, social justice and economic) and weigh these against the arguments that advocate a non-interventionist policy. This will highlight the debate between the Cumulative Causation school of thought and Neo-classical free-market thinkers. The question allows you to come to a conclusion yourself, after you have debated both cases. Of course, examples of government policy need to be used to illustrate your arguments (e.g. the failure of Ravenscraig, or the success of the Welsh Development Agency in attracting inward investment into Glamorgan).

3 (a)   It is very important that only one country in the developed world is considered and you must have a good knowledge of this case study. You must understand what reasons there are for the aided regions' problems, and what reasons the Government has for helping them.

  (b)   The effectiveness of policy will depend upon what goals one sets, and whether comparative differences need to be eliminated, or whether an increase in the absolute wealth of an area (Standard of Living) is simply the goal.

4 (a)   This part of the question is really the same as asking 'Why do governments seek to eliminate regional differences through policy?' That is, for political, social and economic reasons.

  (b)   This second part is concerned with theory, in particular the Myrdal-Freidmann core-periphery model. As with question 2 you also need to cover the argument between free-market thinkers and the CC school. You must remember the question's statement: are there other factors involved, such as inherent or acquired advantages that may also explain imbalanced regional development?

5 (a)   You only need to discuss one of the many models of regional development, such as economic base theory. Perhaps the best may be Myrdal-Freidmann's core-periphery model as it gives you more to discuss than most.

Answers to Practical Geography

Question	Examiner's tip
(b)	This part requires you to use case studies and, as with the other questions in this section, you have to give reasons for spatial inequalities. This may be done best by using one region, such as the Mezzogiorno in Southern Italy, explaining its particular problems (geographical remoteness, harsh climate etc) that are not taken into account by simplistic models. You may, of course, disagree, and argue that the models do explain the differences between regions, but as models they cannot be expected to describe all places but are rather predictions of what is likely to occur. There will always be local reasons in addition to the core-periphery theoretical explanation.

PRACTICAL GEOGRAPHY

Question	Answer	Mark
1		

Examiner's tip The answer should demonstrate the application of geographical techniques, and show a good understanding of how theoretical ideas are applied to problems. This is a straightforward question. You need only follow the instructions and structure given. A mark over 20/25 would get an 'A'.

Microclimate studies are used to determine variations in climate within a few metres of the ground, and within a small area. In this case, the study is of a deciduous wood and its surroundings over the course of a year. Relative humidity, temperature, of both air and soil, and wind characteristics should all be recorded. By conducting the study over such a long period, seasonal variations, rather than diurnal (night-day) differences will be emphasised.

First, you will need a hypothesis, or question, to explore. A possible question is: 'What impact does vegetation have on the microclimate of the area during a year?' Variations may also be caused by changes in albedo (light reflection), aspect and the degree of shelter an area is afforded.

(i) When establishing the transect, some form of spatial sampling must occur i.e. where do you take your readings? This could take the form of a stratified, random or systematic sample. To decide, you must ask 'What information do I need?' The answer is: information from different conditions such as dense wood (this will block out light during the summer when the trees are in leaf), grass and wood margins, as each of these will have different albedos. Therefore, a purposive sample may be the best, as you can consciously choose your sites along the transect; deliberately picking different environmental conditions. This ensures that you get a range of sample environments that random and systematic techniques would not guarantee. The disadvantage of deliberately choosing your sites is that you may subconsciously manipulate the results and cause bias. As the survey is to be carried out over a year, around 4 to 6 sample points would be ideal, more than this and the amounts of data generated would become unmanageable. The sample points should be marked out with something permanent e.g. paint (with the landowner's permission!) so that you return to the same place each time.

(ii) Once the sample points have been selected, readings need to be taken. You would require: wet and dry thermometers or a whirling hygrometer to measure relative humidity, minimum and maximum thermometers to detect diurnal

Answers to Practical Geography

Question	Answer	Mark

variations, and soil thermometers. The wind direction could be recorded by holding a streamer to demonstrate the air flow (using a compass to determine direction) and wind speed could be measured using an anenometer. Ideally, you would have a light meter, to determine incoming radiation and reflection from different vegetation types, although a camera's in-built meter would be a cheaper alternative. Albedo and temperature could then be correlated. Various automatic logging systems are now available with different probes to measure different variables, the information could then be down-loaded onto a computer.

(iii) Data should be collected on a weekly or fortnightly basis, this will provide adequate data, and by adopting a fixed (systematic) sampling period, bias in the data will be avoided.

Once all the information has been collected, statistical analysis (such as Spearman's Rank and chi squared) could be used to explore the hypotheses. For example, is there a relationship between soil/air temperatures and albedo? Are temperatures in the wood always cooler than in the nearby field? By applying statistical tests it is possible to discover whether your hypotheses and expectations were true, false or could have been arrived at through chance. (20)

Question	Examiner's tip

2
- The aim in this question is to give the examiner a clear idea of how you would determine a sphere of influence. It is essential to concentrate on issues such as sampling and difficulties in constructing a questionnaire.

- Draw on any practical work you have done, giving actual examples to back up your points. You might consider a discussion of Reilly's breakpoint and Huff's probability model.

- In the 'Either' part of the question, you need to exclude tourists as you are trying to establish the sphere of influence of the town. You can collect information directly by means of a questionnaire (have as the first question - 'Are you a tourist?'). Consider sampling and timing and location of the survey. You can also collect information indirectly by using delivery areas of shops, secondary school catchment areas and advertising in local newspapers.

- In the 'Or' part of the question, you need to establish where tourists come from. Direct methods for doing this include questionnaires and indirect methods include secondary sources and transport links. The attraction of the settlement to the tourist will be related to its service provision and natural amenities.